T0236782

Communications
in Computer and Information Science   695

Commenced Publication in 2007
Founding and Former Series Editors:
Alfredo Cuzzocrea, Dominik Ślęzak, and Xiaokang Yang

Editorial Board

Simone Diniz Junqueira Barbosa
Pontifical Catholic University of Rio de Janeiro (PUC-Rio),
Rio de Janeiro, Brazil

Phoebe Chen
La Trobe University, Melbourne, Australia

Xiaoyong Du
Renmin University of China, Beijing, China

Joaquim Filipe
Polytechnic Institute of Setúbal, Setúbal, Portugal

Orhun Kara
TÜBİTAK BİLGEM and Middle East Technical University, Ankara, Turkey

Igor Kotenko
St. Petersburg Institute for Informatics and Automation of the Russian
Academy of Sciences, St. Petersburg, Russia

Ting Liu
Harbin Institute of Technology (HIT), Harbin, China

Krishna M. Sivalingam
Indian Institute of Technology Madras, Chennai, India

Takashi Washio
Osaka University, Osaka, Japan

More information about this series at http://www.springer.com/series/7899

Begoña Vitoriano · Greg H. Parlier (Eds.)

Operations Research and Enterprise Systems

5th International Conference, ICORES 2016
Rome, Italy, February 23–25, 2016
Revised Selected Papers

Springer

Editors
Begoña Vitoriano
Complutense University
Madrid
Spain

Greg H. Parlier
INFORMS
Catonsville, MD
USA

ISSN 1865-0929   ISSN 1865-0937 (electronic)
Communications in Computer and Information Science
ISBN 978-3-319-53981-2   ISBN 978-3-319-53982-9 (eBook)
DOI 10.1007/978-3-319-53982-9

Library of Congress Control Number: 2017931543

© Springer International Publishing AG 2017
This work is subject to copyright. All rights are reserved by the Publisher, whether the whole or part of the material is concerned, specifically the rights of translation, reprinting, reuse of illustrations, recitation, broadcasting, reproduction on microfilms or in any other physical way, and transmission or information storage and retrieval, electronic adaptation, computer software, or by similar or dissimilar methodology now known or hereafter developed.
The use of general descriptive names, registered names, trademarks, service marks, etc. in this publication does not imply, even in the absence of a specific statement, that such names are exempt from the relevant protective laws and regulations and therefore free for general use.
The publisher, the authors and the editors are safe to assume that the advice and information in this book are believed to be true and accurate at the date of publication. Neither the publisher nor the authors or the editors give a warranty, express or implied, with respect to the material contained herein or for any errors or omissions that may have been made. The publisher remains neutral with regard to jurisdictional claims in published maps and institutional affiliations.

Printed on acid-free paper

This Springer imprint is published by Springer Nature
The registered company is Springer International Publishing AG
The registered company address is: Gewerbestrasse 11, 6330 Cham, Switzerland

Preface

This text includes extended and revised versions of selected papers from the 5th International Conference on Operations Research and Enterprise Systems (ICORES 2016), held in Rome, Italy during February 23–25, 2016. The conference was organized by the Institute for Systems and Technologies of Information, Control and Communication (INSTICC) and held in cooperation with the Association Française pur la Programmation par Contraintes (AFPC), the Gesellschaft für Operations Research e.V. (GOR) and the Operational Research Society in England (The OR Society). The technical so-sponsor for ICORES is the Portuguese Association of Operational Research (Apdio).

The purpose of the International Conference on Operations Research and Enterprise Systems (ICORES) is to bring together researchers, engineers, and practitioners interested in both theory and applications in the field of operations research. Two simultaneous tracks were held. One track was dedicated to domain-independent methodologies and technologies and another on practical work developed in specific application areas.

ICORES 2016 received 75 paper submissions from 31 countries, across six continents. For each submission, a double-blind review was performed by the Program Committee, whose members are highly qualified researchers in ICORES topic areas. In all, 37 papers were subsequently selected for oral presentation (19 full papers and 18 short papers) and 19 for poster presentation. The full-paper acceptance ratio was about 25%, and the total oral acceptance ratio (including full papers and short papers) was 49%. These competitive acceptance ratios show our intention of preserving a high-quality forum, which we expect to develop further next year.

ICORES 2016 also included three plenary keynote lectures from internationally distinguished researchers: Sue Merchant, (Blue Link Consulting, UK), Alexandre Dolgui, (École des Mines de Nantes, France), and Karla Hoffman, (George Mason University, USA). We express our appreciation for their invaluable contributions and for taking the time to synthesize and prepare their insightful lectures.

We especially thank the authors whose research and development efforts are recorded herein. Finally, we express our gratitude to the entire INSTICC team whose collaboration was fundamental to the success of this productive and memorable conference.

February 2016

Begoña Vitoriano
Greg H. Parlier

Organization

Conference Chair

Dominique de Werra · École Polytechnique Fédérale de Lausanne (EPFL), Switzerland

Program Co-chairs

Begoña Vitoriano · Complutense University, Spain
Greg H. Parlier · INFORMS, USA

Program Committee

El-Houssaine Aghezzaf · Ghent University, Belgium
Javier Alcaraz · Universidad Miguel Hernandez de Elche, Spain
Maria Teresa Almeida · ISEG, Universidade de Lisboa, Portugal
Lionel Amodeo · University of Technology of Troyes, France
Necati Aras · Bogazici University, Turkey
Lyes Benyoucef · Aix-Marseille University, France
Endre Boros · Rutgers University, USA
Ahmed Bufardi · None, Switzerland
Sujin Bureerat · KhonKaen University, Thailand
Giuseppe Buttazzo · University of Pisa, Italy
Alfonso Mateos Caballero · Universidad Politécnica de Madrid, Spain
José Manuel Vasconcelos Valério de Carvalho · Universidade do Minho, Portugal
Bo Chen · University of Warwick, UK
John Chinneck · Carleton University, Canada
James Cochran · University of Alabama, USA
Mikael Collan · Lappeenranta University of Technology, Finland
Ademir Constantino · Universidade Estadual de Maringá, Brazil
Alysson Costa · University of Melbourne, Australia
Paolo Dell'Olmo · Sapienza University of Rome, Italy
Xavier Delorme · Ecole Nationale Supérieure des Mines de Saint-Etienne, France
Marc Demange · RMIT University, Australia
Clarisse Dhaenens · French National Institute for Research in Computer Science and Control, France
Tadashi Dohi · Hiroshima University, Japan
Nikolai Dokuchaev · Curtin University, Australia

Gintautas Dzemyda	Vilnius University, Lithuania
Andrew Eberhard	RMIT University, Australia
Khaled Elbassioni	Masdar Institute, United Arab Emirates
Nesim Erkip	Bilkent University, Turkey
Laureano F. Escudero	Universidad Rey Juan Carlos, Spain
Gerd Finke	Grenoble Institute of Technology—Lab. G-SCOP, France
Christodoulos Floudas	Texas A&M University, USA
Muhammad Marwan Muhammad Fuad	Aarhus University, Denmark
Robert Fullér	Obuda University, Hungary
Ron Giachetti	Naval Postgraduate School, USA
Giorgio Gnecco	IMT, School for Advanced Studies, Lucca, Italy
Marc Goerigk	Lancaster University, UK
Boris Goldengorin	Ohio University, USA
Marta Castilho Gomes	CERIS-CESUR, Instituto Superior Técnico, Universidade de Lisboa, Portugal
Juan José Salazar Gonzalez	Universidad de La Laguna, Spain
Dries Goossens	Ghent University, Belgium
Nalan Gulpinar	University of Warwick, UK
Gregory Z. Gutin	Royal Holloway University of London, UK
Jin-Kao Hao	University of Angers, France
Hanno Hildmann	Universidad Carlos III de Madrid, Spain
Han Hoogeveen	Universiteit Utrecht, The Netherlands
Chenyi Hu	University of Central Arkansas, USA
Johann Hurink	University of Twente, The Netherlands
Josef Jablonsky	University of Economics, Czech Republic
Itir Karaesmen	American University, USA
Abdelhakim Khatab	Lorraine University, France
Jesuk Ko	Gwangju University, Republic of Korea
Leszek Koszalka	Wroclaw University of Technology, Poland
Philippe Lacomme	Université Clermont-Ferrand 2 Blaise Pascal, France
Sotiria Lampoudi	Liquid Robotics Inc., USA
Dario Landa-Silva	University of Nottingham, UK
Pierre L'Ecuyer	Université de Montreal, Canada
Janny Leung	The Chinese University of Hong Kong (Shenzhen), China
Abdel Lisser	The University of Paris-Sud 11, France
Pierre Lopez	LAAS-CNRS, Université de Toulouse, France
Manuel López-Ibáñez	University of Manchester, UK
Helena Ramalhinho Lourenço	Universitat Pompeu Fabra, Spain
Prabhat Mahanti	University of New Brunswick, Canada
Viliam Makis	University of Toronto, Canadá
Concepción Maroto	Universidad Politécnica de Valencia, Spain

Pedro Coimbra Martins	Polytechnic Institute of Coimbra, Portugal
Nimrod Megiddo	IBM Almaden Research Center, USA
Carlo Meloni	Politecnico di Bari, Italy
Marta Mesquita	Universidade de Lisboa, Portugal
Rym M. Hallah	Kuwait University, Kuwait
Michele Monaci	Università degli Studi di Bologna, Italy
Jairo R. Montoya-Torres	Universidad de los Andes, Colombia
Young Moon	Syracuse University, USA
Stefan Nickel	Karlsruhe Institute of Technology (KIT), Germany
Gaia Nicosia	Università degli Studi Roma Tre, Italy
José Oliveira	Universidade do Minho, Portugal
Selin Özpeynirci	Izmir University of Economics, Turkey
Andrea Pacifici	Università di Roma Tor Vergata, Italy
Greg H. Parlier	INFORMS, USA
Sophie Parragh	University of Vienna, Austria
Vangelis Paschos	University of Paris-Dauphine, France
Ulrich Pferschy	University of Graz, Austria
Diogo Pinheiro	Brooklyn College of the City University of New York, USA
Ed Pohl	University of Arkansas, USA
Steven Prestwich	University College Cork, Ireland
Arash Rafiey	Indiana State University, USA
Günther Raidl	Vienna University of Technology, Austria
Martín Gomez Ravetti	Universidade Federal de Minas Gerais, Brazil
Celso Ribeiro	Universidade Federal Fluminense, Brazil
Michela Robba	University of Genoa, Italy
Andre Rossi	Université d'Angers, France
Stefan Ruzika	University of Koblenz-Landau, Germany
Marcello Sanguineti	University of Genoa, Italy
Cem Saydam	University of North Carolina Charlotte, USA
Andrea Schaerf	Università di Udine, Italy
Rene Seguin	Defence Research and Development Canada, Canada
Marc Sevaux	Université de Bretagne-Sud, France
Thomas Stützle	Université Libre de Bruxelles, Belgium
Jacques Teghem	Faculté Polytechnique de Mons, Belgium
Vadim Timkovski	South University, USA
Norbert Trautmann	University of Bern, Switzerland
Alexis Tsoukiàs	CNRS, France
Begoña Vitoriano	Complutense University, Spain
Maria Vlasiou	Eindhoven University of Technology, Netherlands
Luk N. Van Wassenhove	INSEAD Business School, France
Dominique de Werra	École Polytechnique Fédérale de Lausanne (EPFL), Switzerland
Santoso Wibowo	CQ University, Australia
Marino Widmer	University of Fribourg, Switzerland
Gerhard Woeginger	Eindhoven University of Technology, The Netherlands

Riccardo Zecchina   Politecnico di Torino, Italy
Hongzhong Zhang   Columbia University, USA
Sanming Zhou   University of Melbourne, Australia

Additional Reviewers

Fabian Dunke   KIT, Germany
Francisco Saldanha   University of Lisbon, Portugal
  da Gama
Gustavo Campos Meneses   Centro Federal de Educação Tecnológica de Minas
                               Gerais, Brazil
André Gustavo dos Santos   Universidade Federal de Viçosa, Brazil

Invited Speakers

Sue Merchant   Blue Link Consulting, UK
Alexandre Dolgui   École des Mines de Nantes, France
Karla Hoffman   George Mason University, USA

Contents

Methodologies and Technologies

An Investigation on Compound Neighborhoods for VRPTW

Binhui Chen[1]([✉]), Rong Qu[1], Ruibin Bai[2], and Hisao Ishibuchi[3]

[1] University of Nottingham, Nottingham, UK
{bxc,rxq}@cs.nott.ac.uk
[2] University of Nottingham Ningbo, Ningbo, China
ruibin.bai@nottingham.edu.cn
[3] Osaka Prefecture University, Osaka, Japan
hisaoi@cs.osakafu-u.ac.jp

Abstract. The Vehicle Routing Problem with Time Windows (VRPTW) consists of constructing least cost routes from a depot to a set of geographically scattered service points and back to the depot, satisfying service time intervals and capacity constraints. A Variable Neighbourhood Search algorithm which can simultaneously optimize both objectives of VRPTW (to minimize the number of vehicles and the total travel distance) is proposed in this paper. The three compound neighbourhood operators are developed with regards to problem characteristics of VRPTW. Compound neighbourhoods combine a number of independent neighbourhood operators to explore a larger scale of neighbourhood search space. Performance of these operators has been investigated and is evaluated on benchmark problems.

Keywords: Variable Neighbourhood Search · Vehicle Routing Problem with Time Windows · Compound neighbourhood · Metaheuristics

1 Introduction

The Vehicle Routing Problem (VRP) [1] is an important transport scheduling problem which can be used to model various real-life problems, such as postal deliveries, school bus routing, recycling routing and so on.

1.1 Problem Description and Related Work

The Vehicle Routing Problem with Time Windows (VPRTW) is a classical variant of VRP, which can be defined as follows. Let $G = (V, E)$ be a directed graph where $V = \{v_i, \ i = 0, \ldots, n\}$ denotes a depot (v_0) and n customers ($v_i, \ i = 1, \ldots, n$). A non-negative service demand q_i and service time s_i are associated with v_i, while $q_0 = 0$ and $s_0 = 0$. E is a set of edges with non-negative weights d_{ij} (which often represents distance) between v_i and v_j ($v_i, v_j \in V$).

All customer demands are served by a fleet of K vehicles. To customer v_i, the service start time b_i must be in a time window $[e_i, f_i]$, where e_i and f_i

© Springer International Publishing AG 2017
B. Vitoriano and G.H. Parlier (Eds.): ICORES 2016, CCIS 695, pp. 3–19, 2017.
DOI: 10.1007/978-3-319-53982-9_1

are the earliest and latest time to serve q_i. If a vehicle arrives at v_i at time $a_i < e_i$, a waiting time $w_i = max\{0, e_i - a_i\}$ is required. Consequently, the service start time $b_i = max\{e_i, a_i\}$. Each vehicle of a capacity Q travels on a route connecting a subset of customers starting from v_0 and ending within schedule horizon $[e_0, f_0]$. The decision variable $X_{ij}^k = 1$ if the edge from v_i to v_j is assigned in route k ($k \in K$); Otherwise $X_{ij}^k = 0$. The objective functions can be defined as follows [2]:

$$Minimize \qquad K \qquad (1)$$

$$Minimize \qquad \sum_{k \in K} \sum_{v_i \in V} \sum_{v_j \in V} X_{ij}^k \cdot d_{ij} \qquad (2)$$

Subject to:

$$\sum_{k \in K} \sum_{v_j \in V} X_{ij}^k = 1 \qquad \forall v_i \in V \backslash \{v_0\} \qquad (3)$$

$$\sum_{k \in K} \sum_{v_i \in V} X_{ij}^k = 1 \qquad \forall v_j \in V \backslash \{v_0\} \qquad (4)$$

$$\sum_{k \in K} \sum_{v_i \in V} \sum_{v_j \in V \backslash \{v_0\}} X_{ij}^k = n \qquad (5)$$

$$\sum_{v_j \in V} X_{0j}^k = 1 \qquad \forall k \in K \qquad (6)$$

$$\sum_{v_i \in V} X_{ij}^k - \sum_{v_i \in V} X_{ji}^k = 0 \qquad \forall k \in K, v_j \in V \backslash \{v_0\} \qquad (7)$$

$$\sum_{v_i \in V} X_{i0}^k = 1 \qquad \forall k \in K \qquad (8)$$

$$e_i \leq b_i \leq f_i \qquad \forall v_i \in V \qquad (9)$$

$$\sum_{v_i \in V} \sum_{v_j \in V} X_{ij}^k \cdot q_i \leq Q \qquad \forall k \in K \qquad (10)$$

$$X_{ij}^k \in \{0, 1\} \qquad \forall v_i, v_j \in V, k \in K \qquad (11)$$

Objective (1) aims to minimize the requested number of vehicles. Objective (2) minimizes the total travel distance of the fleet. Constraints (3)–(5) limit every customer to be served exactly once and all customers are visited. Constraints (6)–(8) define the route by vehicle k. Constraints (9) and (10) guarantee the feasibility with respect to the time constraints on service demands and capacity constraints (Q) on vehicles, respectively. Constraint (11) defines the domain of the decision variable X_{ij}^k.

Most researchers consider minimizing the number of vehicles as the primary objective [3], while others study it as a multi-objective problem [4]. In the former case, a two-phase approach is often used, to minimize the vehicle number firstly and then minimize the distance with a fixed route number in the second phase. Population-based methods are usually used for solving the multi-objective

VRPTW. Other objectives in VRPTW include the minimization of the total waiting time and so on, which are less studied [5,6].

Due to the problem size and NP-hard property of VRPTW, standard mathematical methods often perform poorly within a reasonable amount of time [1]. Metaheuristics and hybrid algorithms have attracted more attention in VRPTW. They could be grouped into *population-based* metaheuristics and *local search* metaheuristics. Population-based methods work on a set of candidate solutions which requires a high computation cost. This is a main drawback for them to achieve high performance in VRP. More details could be found in [7].

Many local search approaches have been applied to VRPTW, such as Tabu Search [8], Simulated Annealing [9] and Variable Neighbourhood Search (VNS) [10]. This paper focuses on VNS methods. Its first application is on TSP with and without backhaul [11].

VNS shifts among different neighbourhood structures which define different search spaces. Different variants of VNS have been studied in the literature. In the Basic VNS, a *Local Search* finds local optimal solutions using different neighbourhood structures, and *Shaking* is used to perturb the search to enhance diversification. Variable Neighbourhood Descent (VND) algorithm changes the neighbourhoods in a deterministic way [12]. Reduced VNS [13] selects neighbourhood moves randomly from a neighbourhood set. General VNS [14] is an extension of Basic VNS, whose the local search is a VND as well.

VNS and its extensions have been studied extensively in various VRP problems. Literature [3] proposes a four-phase approach based on VND for VRPTW, while literature [15] develops a VNS for Multi-Depot Vehicle Routing Problem with Time Windows where routes start and end at different depots. A VNS algorithm for the Open Vehicle Routing Problem without time constraint is presented in [16]. The study in [17] concentrates on the Periodic Vehicle Routing Problem, where the schedule horizon is very large without time constraint. An extensive review on VNS can be found in [10].

1.2 Widely Used Neighbourhood Operators in VRP

Neighbourhood operators define the search spaces of different features, thus significantly affect the success of local search. Neighbourhood moves in VRP can be classified into two categories: *Inter-Route* exchange and *Intra-Route* exchange (some authors use the term *interchange* instead of *exchange*), which exchange nodes or edges among routes or within one route, respectively.

Literature [18] introduces the λ-*optimality* mechanism, which is widely applied in routing problems. It removes λ edges from one route, and reconnects it in a feasible way. *2-opt* and *3-opt* are two typical operators of this mechanism, both may reverse the order of nodes. *Or-opt* [19] is a specific subset of *3-opt* operators, and it includes only those moves which do not reverse customer links. The study in [20] introduces the λ-*interchange* mechanism which exchanges two groups of nodes from different routes. The number of nodes in each group should not be more than λ, while the nodes are not necessarily consecutive.

In *CROSS-exchange* [21], two strings of consecutive nodes from two routes are exchanged, preserving the order of customers in each string.

The above VNS approaches use *independent moves* in each single neighbourhood operator. The approach in [22] combines independent moves such as 2-opts, swaps and insertions in a very large scale neighbourhood search. This method is applied to TSP and VRP with side constraints of capacity and distance. The independent moves in this method are different on the operation position while their operator settings are the same. The study shows that this kind of *compounded neighbourhoods* are competitive for solving VRP. This kind of compounding method with sequential addition and deletion of edges is also used by Ejection Chain approach [23]. In the next section, we propose and study compound neighbourhoods in a different compounding way for VRPTW.

2   Variable Neighbourhood Search with Compound Neighbourhoods

We propose three compound neighbourhood operators within the General VNS (VNS-C) in this research. In our approach, compared to existing neighbourhood operators, the independent operators not only operate on different positions within the routes, but also have different lengths of exchange segments. A deterministic constraint is imposed to the lengths of the exchange segments, which are selected with a random selection scheme. By using this way of compounding operation upon both intra-route and inter-route neighbourhood solutions, two compound neighbourhoods are produced. In addition, a third neighbourhood operator which compounds segment insertion operators with the same length limit is also developed to reduce vehicle number. By adopting these operators, VNS-C optimizes both objectives *simultaneously* in each iteration by shifting among broader search regions.

2.1   Compound Operators and Neighbourhoods

Because of the time window constraint in VRPTW, the reverse operation in the standard λ-opt and λ-interchange operators tends to bring infeasibility, thus *Or-opt* and *CROSS-exchange* are adopted in the *Compound Operators* in VNS-C.

Based on the *Or-opt* exchange operator, we devise an improved *Or-opt* intra-route operator *Or-opt-i*, where i is the length limit of two randomly selected exchange links. In an *Or-opt-i* exchange, the length of one exchange link is fixed to i to avoid redundant exchanges while the length of the other exchange link is randomly set to up to i. This exchange link length setting cooperates with random operation position selection, composing the compounding manner of the proposed compound neighbourhoods. For example, *Or-opt-3* is a compound neighbourhood which assembles three independent neighbourhood moves (where one exchange link length is fixed to 3, while the other one's length could be 1, 2 or 3). An illustrative example is presented in Fig. 1 (a), where directions of both operated links are kept.

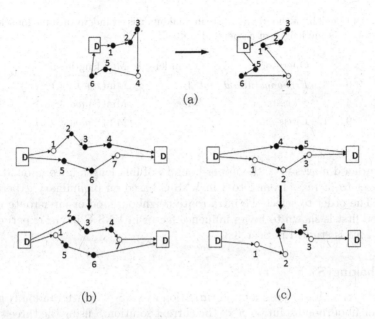

Fig. 1. Examples of neighbourhoods in VNS-C. (a). Link (1,2,3) exchanged with link (5,6) by the intra-route *Or-opt-i* compound operator (*i*=3). (b). Link (2,3,4) exchanged with link (5,6) by the inter-route compound operator *CROSS-i* (*i*=3). (c). Link (4,5) is removed by *LinkMove-i* from its original route and inserted to a random position in the target route, which brings a route reduction.

A *CROSS-i* compound operator is also proposed in VNS-C. It includes all independent *CROSS*-exchanges to exchange a link of length *i* with another link of length up to *i* between routes. For instance, the compound neighbourhood of *CROSS-3* assembles three independent neighbourhood moves where length of one exchange link is fixed to 3 and the other one is randomly set to up to 3 (denoted as 3-1, 3-2 and 3-3 independent exchanges, respectively). In one move of *CROSS-3*, the best solution among all the three independent neighbourhoods is selected. While, in a standard independent neighbourhood search, the best solution based on only one of the 3-1, 3-2 or 3-3 exchanges will be selected. An example of *CROSS-3* is presented in Fig. 1 (b), where the selected improvement solution is produced by a 3-2 exchange.

In the proposed VNS-C, an operator named *LinkMove-i* is developed to reduce both the vehicle number and total travel distance simultaneously (*i* is the max length of operated links), rather than in two separate phases. In *LinkMove-i*, a customer link of length α ($\alpha \le i$) from route h is removed and reinserted into route t ($h \ne t$). When α is equal to the length of route h, route h would be removed thus leads to a solution with one less route. Fig. 1 (c) presents an example of *LinkMove-i*.

The proposed compound neighbourhoods explore larger search areas than standard independent neighbourhoods. Following the rule of invoking small

Table 1. Set of neighbourhood operators in Shaking. z is a random variable for selecting an operator. L is the length of associated routes.

z	Operator	Min length	Max length(ml)
$0 \sim 3$	*ExchangeInRoute-ml*	1	Min($z + 1$, L)
$4 \sim 8$	*Move-ml*	1	Min($z \bmod 3$, L)
$9 \sim 12$	*Cross-ml*	1	Min($z \bmod 8$, L)

neighbourhood moves first, the upper bound i of link length in compound operators is set to increase from 1 to 5 in VNS-C based on preliminary experiment results. The order to select the intra-route neighbourhood or inter-route neighbourhood first is shown to be an influence factor in VNS-C in our experimental results. This is studied in Sect. 3.3.

2.2 Shaking(S)

Shaking(S) is a phase of random perturbation in VNS-C, which randomly generates a neighbourhood solution S' of the current solution S using the three simple operators in Table 1, aiming to escape from local optima. *ExchangeInRoute-ml* and *Cross-ml* exchanges two segments within one route and between two randomly selected routes, respectively. *Move-ml* inserts a randomly selected route segment from a route to another route. In all three operators, the maximum length of the segment is ml. Different from the above-mentioned compound operators, operators in *Shaking* are more flexible, without the requirement that at least one segment's length must be ml.

The first feasible move will be accepted in *Shaking(S)*. To encourage farther moves, the segment of length ml is selected with a higher probability. If no feasible moves are found after a pre-specified number of evaluations, the original input solution S would be returned. We investigate this process in Sect. 3.2.

2.3 Local Search

In the local search of VNS-C (see *Algorithm* 1), NS_{max} evaluations are undertaken in each run of neighbourhoods. Literature [10] recommends that, when the initial solution is constructed by a heuristic, the Best-Improvement acceptance criterion should be used in VNS. The initial solution in VNS-C is constructed using the Nearest Neighbourhood heuristic from [6], and the best neighbourhood solutions are chosen. To avoid being stuck to local optima, *Record-to-Record Travel* algorithm [24] is adopted as the acceptance criteria, where *Quality()* is defined by the total travel distance, and *DEVIATION* is set to 15. Here a solution with the lower *Quality()* value is better. The search stops at a time limit of $Time_{max}$ or when all three compound neighbourhoods are estimated but no improvement is found. In *Algorithm* 1, $N_r(S', i)$ represents the rth neighbourhood operator applied to the incumbent solution S' with operated link length limit of i.

2.4   The VNS-C Framework

The pseudo-code of VNS-C is presented in *Algorithm* 2, where the iteration time is set to C_{max}. In *Step 1*, an initial solution is constructed using a heuristic, which inserts the "closest" available customer into the incumbent partial route. Here the distance between two customers is defined by their Geographic distance d_{ij}, Temporal distance T_{ij} and the degree of Emergency v_{ij} which are used in [6], shown in (12) as below:

$$Dis = \delta_1 \cdot d_{ij} + \delta_2 \cdot T_{ij} + \delta_3 \cdot v_{ij} \quad s.t. \ \delta_1 + \delta_2 + \delta_3 \ = \ 1 \qquad (12)$$

The three coefficients δ_1, δ_2 and δ_3 define the importance of each component in the distance definition. We set them as $\delta_1 = 0.4$, $\delta_2 = 0.4$ and $\delta_3 = 0.2$ (empirically calculated by [6]).

3   Experiments

3.1   Problem Dataset and Parameter Setting

The proposed VNS-C was evaluated on the Solomon benchmark [6], which consists of six datasets (C1, C2, R1, R2, RC1, RC2), each has eight to 12 instances of 100 customers with their own service demands. In C1 and C2, customers are located in a number of clusters, while the objectives of (1) and (2) are positively related [4]. Customers of R1 and R2 are randomly distributed geographically, while RC1 and RC2 are a mix of them. The scheduling horizons in C1, R1 and RC1 are short, and their vehicle capacities are low (200). C2, R2 and RC2 have higher vehicle capacities (700, 1000 and 1000, respectively), leading to fewer required vehicles to satisfy all demands. Diverse time window widths are distributed with various densities.

Tuning is conducted on only one parameter at a time, while fixing all the others on a small number of instances. Preliminary experiments show that most feasible solutions in Shaking are found in around 200 evaluations, thus 300 evaluations are conducted. For each incumbent solution in the local search, 400 neighbourhoods are evaluated, i.e. $NS_{max} = 400$. $Time_{max}$ is set to 1,000,000 evaluations while the max iteration time C_{max} of VNS-C is 300. All results are produced in 30 runs to conduct statistical analysis.

3.2   Compound Neighbourhoods and Shaking

Table 2 presents the average results from VNS-C, VNS with independent operators (standard Or-opt and CROSS exchange) and VNS-C without Shaking on six randomly chosen instances. NV denotes the number of vehicles, TD represents the total travel distance, and Times is the total number of evaluations. S.D is the standard deviation. It is shown that VNS-C produces significantly better and more stable results compared to the other variants. *Shaking* also improves VNS-C in terms of both quality and stability, thus is an essential and necessary component in VNS-C.

Algorithm 1. Local Search(S', S).

Step 1: Input solution S' and S.
Step 2: Set $r \leftarrow 1$, $i \leftarrow 1$, $time \leftarrow 0$.
    while ($r < 4$ **And** $time < Time_{max}$) **do**
        Step 2.1: Neighbourhood Search
            $S'' \leftarrow$ Best Improvement of $N_r(S', i)$.
            $time \leftarrow time + NS_{max}$.
        Step 2.2: Move or Not
            if $Quality(S'') < Quality(S)$ **then**
                $S \leftarrow S''$, $S' \leftarrow S''$, $i \leftarrow 0$, $r \leftarrow 1$.
            else if $Quality(S'') - Quality(S) <$
            $DEVIATION$ **then**
                $S' \leftarrow S''$, $i \leftarrow 0$, $r \leftarrow 1$.
            end if
        Step 2.3: Shift Neighbourhood Structure
            $i \leftarrow i + 1$.
            if $i = 6$ **then** $r \leftarrow r + 1$, $i \leftarrow 1$.
    end while
Step 3: Output the best found solution S.

Algorithm 2. The VNS-C framework.

Step 1: Generate an initial feasible solution S by the Nearest Neighbourhood heuristic.
Step 2:
    Set $C \leftarrow 1$.
    while $C < C_{max}$ **do**
        Step 2.1: $S' \leftarrow$ **Shaking**(S).
        Step 2.2: $S \leftarrow$ **Local Search**(S', S).
        Step 2.3:
            if S is improved **then**
                $C \leftarrow 1$.
            else
                $C \leftarrow C + 1$.
            end if
    end while
Step 3: Output S.

To verify whether the result of VNS-C is *significantly different* from the other three algorithms', T-test is executed between results of VNS-C and the other three algorithms. Here the confidence level is set as 95%. Table 3 presents the test result, where Y represents two populations are significantly different, and N the opposite. Notably, as the solutions have two dimensions of NV and TD, as long as the p-value (two-tail) is smaller than 5% in one dimension, the two populations would be considered as significantly different. It can be seen that VNS-C produces significantly better solutions than the other three algorithms on complicated instances (R and RC). For the two C instances, there is no significant difference between VNS-C and the other three algorithms. Results against those

Table 2. Comparison of VNS-C and VNS of Independent Operator with and without Shaking. Best results are in bold.

Instance			C101	C201	R101	R201	RC101	RC201
VNS-C &	Best	NV	10	3	19	4	15	4
Shaking		TD	828.94	591.56	1643.34	1190.52	1624.97	1310.44
	Average	NV	10	3	19.9	4.83	15.6	4.93
		TD	828.94	591.56	1647.9	1246.91	1652.38	1365.76
		Times	67,247,717	97,011,547	188,429,463	176,349,146	113,982,405	137,842,632
	S.D on NV		0	0	0.31	0.38	0.63	0.25
	S.D on TD		0	0	5.59	45.12	12.88	41.57
Independent	Best	NV	10	3	19	4	16	4
Operators &		TD	828.94	591.56	1700.42	1339.84	1753.49	1482.86
Shaking	Average	NV	10	3	20.13	4.73	16.3	4.97
		TD	828.94	591.56	1791.73	1538.66	1878.68	1569.51
		Times	9,808,756	9,800,000	159,166,298	128,370,863	79,996,218	146,881,580
	S.D on NV		0	0	0.68	0.45	0.47	0.18
	S.D on TD		4.38	0	89.32	212.51	101.68	59.89
VNS-C	Best	NV	10	3	19	4	15	4
		TD	828.94	591.56	1644.55	1294.36	1644.18	1340.79
without	Average	NV	10	3	20.43	4.73	16.43	4.93
Shaking		TD	828.94	591.56	1823.72	1511.68	1856.99	1489.35
		Times	10,026,352	97,010,797	77,128,988	124,141,441	104,983,387	113,826,219
	S.D on NV		0	0	0.63	0.45	0.68	0.25
	S.D on TD		0	0	178.68	394.21	213.67	270.99
Independent	Best	NV	10	3	19	4	15	4
Operators		TD	828.94	591.56	1649.23	1226.43	1624.97	1332.74
without	Average	NV	10	3	20.33	4.6	16.17	4.97
Shaking		TD	828.94	591.56	1828.49	1671.69	1859.35	1443.03
		Times	10,198,465	97,066,537	24,214,377	101,344,273	60,738,070	128,573,526
	S.D on NV		0	0	0.76	0.5	0.91	0.18
	S.D on TD		0	0	173.87	401.52	211.84	199.56

Table 3. T-test between VNS-C and the other three algorithms.

Compared Algorithms		C101	C201	R101	R201	RC101	RC201
Independent	P-value(NV)	1	1	0.094698	0.355754	1.49E-06	0.561629
Neighbourhoods	P-value(TD)	0.321464	1	1.1E-09	2.3E-08	1.33E-05	6.62E-21
with Shaking	Different	N	N	Y	Y	Y	Y
Compound	P-value(NV)	1	1	0.000138	0.355754	0.004581	1
Neighbourhoods	P-value(TD)	1	1	8.66E-06	0.000976	9.92E-06	0.019461
without Shaking	Different	N	N	Y	Y	Y	Y
Independent	P-value(NV)	1	1	0.006101	0.046114	5.73E-07	0.561629
Neighbourhoods	P-value(TD)	1	1	3.78E-06	2.74E-06	4.91E-13	0.04599
without Shaking	Different	N	N	Y	Y	Y	Y

in the literature (see Table 5) indicate that all these four algorithms obtained the best solution for these two instances, thus no significant difference has been found.

Table 4. Results of VNS-C with four different operator orders. Best results are in bold.

Instance			C101	C201	R101	R201	RC101	RC201
MCI	Best	NV	10	3	19	4	15	4
		TD	**828.94**	**591.56**	1643.34	**1190.52**	1624.97	**1310.44**
	Average	NV	10	3	19.9	**4.83**	**15.6**	**4.93**
		TD	**828.94**	**591.56**	1647.9	**1246.91**	1652.38	**1365.76**
		Times	67,247,717	97,011,54	188,429,463	176,349,146	113,982,405	137,842,632
CMI	Best	NV	10	3	20	5	15	5
		TD	**828.94**	**591.56**	**1642.88**	**1189.82**	**1623.58**	1317.97
	Average	NV	10	3	20.3	5	15.87	5
		TD	**828.94**	**591.56**	1650.42	**1214.66**	1654.65	1367.59
		Times	69,224,764	97,011,573	181,014,272	166,688,866	184,803,844	158,724,815
ICM	Best	NV	10	3	20	4	15	5
		TD	**828.94**	**591.56**	**1642.88**	1237.76	1715.49	1354.72
	Average	NV	10	3	20.17	4.9	16.4	5
		TD	**828.94**	**591.56**	1648.9	1306.7	1762.34	1425.46
		Times	51,225,270	97,015,461	223,733,841	307,952,033	692,934,814	516,816,576
IMC	Best	NV	10	3	19	4	16	5
		TD	**828.94**	**591.56**	1643.18	1234.09	1672.33	1376.17
	Average	NV	10	3	**19.9**	**4.7**	16.27	5
		TD	**828.94**	**591.56**	**1647.52**	**1334.32**	1765	1464.56
		Times	67,357,647	97,015,347	199,078,925	311,452,951	735,044,303	492,990,103

3.3 Neighbourhoods Order

Table 4 compares different orders of neighbourhoods employed in VNS-C (M, C and I represent *LinkMove-i*, *CROSS-i* and *Or-opt-i*, respectively). It can be seen that, the Inter-Route move first group (MCI and CMI) achieves better results than the Intra-Route move first ones (ICM and IMC). In the former case, MCI performs better than CMI. It seems that optimizing the route number first, and then assigning customers to a route and optimizing the customer order in each route can bring better results. On R101 and R201, MCI obtains better NV while CMI has better TD. As objective (1) is usually considered as primary, the order of MCI will be used in VNS-C.

3.4 Experiment Results and Analysis

Table 5 presents the results on all the 56 Benchmark Solomons instances. It illustrates that VNS-C is effective in improving both objectives simultaneously. In problems whose objectives are positively correlated, VNS-C can produce the current best known solutions in a reasonable time. In other instances, some better solutions with less TD are found comparing to the best known solutions with the same NV in the literature.

It is also shown that VNS-C is effective in minimizing TD by the results on the complicated datasets (R2, RC2) which use fewer vehicles to satisfy 100 demands. Thus these instances require a powerful neighbourhood operator to reduce NV. Figure 2 shows that the disparity between our best found solution (NV = 3 and TD = 766.91) and the best known solution with a lower NV (NV = 2 and TD = 825.52) on R204 is large, which indicates that the distance between them is large in the search space. It is difficult for VNS-C to find a lower NV in this case, as the link length limit in *LinkMove-i* (5) is too small compared to the route length (33).

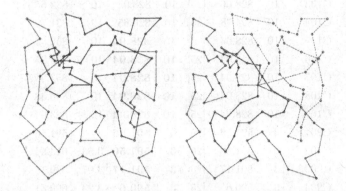

Fig. 2. Two best found solutions with 2 and 3 vehicles on R204.

To investigate the performance of *LinkMove-i* on reducing NV, six different upper bound values of i (Max_i) are set to this operator. Figures 3 and 4 demonstrate the performance of the *LinkMove-i* operator with diverse Max_i on minimizing NV. In Fig. 3 it can be seen that, higher Max_i can produce a lower average NV on four complicated instances (R101, R201, RC101 and R201) while the best found NV is unchanged. Too small Max_i would insert only short routes to other routes, and the capability of reducing NV would become weaker consequently. This hypothesis is consistent to the observation on RC201 in Fig. 4 that, when Max_i is small (1, 2 and 3) the best found NV (5) is greater than the one (4) found with larger Max_i values (5, 7 and 9). In addition, experiment results also show that some of the lowest TD can be found with small Max_i, while their average NVs are the largest. e.g. the lowest TDs are found on R101, RC101 and RC201 with $Max\_i = 1$, as well as on R201 with $Max\_i = 2$. This observation indicates the conflicting relation of both objectives on these instances.

Or-opt-i and *CROSS-i* reduce TD by intra-route moves and inter-route moves respectively. In theory, the neighbourhoods of *CROSS-i* cover the neighbourhoods of *Or-opt-i* in search space, while more move steps are required. To clarify whether *Or-opt-i* is redundant in our algorithm and which operator makes more contribution on reducing TD, we respectively remove one of these two operators from VNS-C and evaluate the two algorithm variants. The experiment results are presented in Table 6, which gives the gaps between the new variants' results and

Table 5. VNS-C on Benchmark Solomon's instances. Best found solutions in the literature are shown in bold.

Instance	Best Known			VNS-C			
	NV	TD	Ref.	Best		Average	
				NV	TD	NV	TD
C101	10	828.94	[25]	**10**	**828.94**	10	828.94
C102	10	828.94	[25]	**10**	**828.94**	10	876.79
C103	10	828.06	[25]	10	828.94	10	832.65
C104	10	824.78	[25]	10	825.65	10	831.79
C105	10	828.94	[25]	**10**	**828.94**	10	852.33
C106	10	828.94	[25]	**10**	**828.94**	10.07	836.25
C107	10	828.94	[25]	**10**	**828.94**	10	853.9
C108	10	828.94	[25]	**10**	**828.94**	10	840.48
C109	10	828.94	[25]	**10**	**828.94**	10	823.94
C201	3	591.56	[25]	**3**	**591.56**	3	591.56
C202	3	591.56	[25]	**3**	**591.56**	3.53	613.94
C203	3	591.17	[25]	**3**	**591.17**	3.07	599.16
C204	3	590.6	[25]	**3**	**590.6**	3.23	609.81
C205	3	588.88	[25]	**3**	**588.88**	3	588.88
C206	3	588.49	[25]	**3**	**588.49**	3	588.49
C207	3	588.29	[25]	**3**	**588.29**	3	588.29
C208	3	588.32	[25]	**3**	**588.32**	3	588.32
R101	19	1650.80	[25]	19	1652.47	19.9	1647.90
	20	1642.87	[26]	20	1643.34		
R102	17	1486.12	[25]	18	1476.06	18.9	1493.30
R103	13	1292.67	[25]	14	1219.89	14.17	1230.92
R104	9	1007.31	[25]			11.1	1009.9
	10	974.24	[27]	10	1007.27		
	11	971.5	[28]	11	994.85		
R105	14	1377.11	[25]	14	1381.88	15.07	1377.24
	15	1346.12	[29]	15	1360.78		
R106	12	1252.03	[25]			13.57	1264.04
	13	1234.6	[30]	13	1243.72		
R107	10	1104.66	[25]			11.73	1097.07
	11	1051.84	[29]	11	1077.24		
R108	9	960.88	[25]			10.23	974.46
	10	932.1	[31]	10	956.22		
R109	11	1194.73	[25]			12.93	1181.99
	12	1013.2	[32]	12	1168.18		
	13	1151.84	[26]	13	1157.61		

Table 5. *(Continued)*

Instance	Best Known			VNS-C			
	NV	TD	Ref.	Best		Average	
				NV	TD	NV	TD
R110	10	1118.84	[25]			12.1	1106.02
	11	1112.21	[31]				
	12	1068	[30]	12	1081.88		
R111	10	1096.72	[25]	**11**	**1087.5**	11.9	1080.1
	12	1048.7	[30]	12	1062.58		
R112	9	982.14	[25]			10.9	979.52
	10	953.63	[33]	10	958.7		
R201	4	1252.37	[25]	4	1282.75	4.83	1246.91
	5	1206.42	[27]	**5**	**1190.52**		
R202	3	1191.7	[25]			4	1146.34
	4	1091.21	[27]	4	1098.06		
R203	3	939.503	[25]	3	968.67	3.5	969.05
	4	935.04	[27]	**4**	**905.72**		
R204	2	825.52	[25]			3	809.88
	3	789.72	[27]	**3**	**766.91**		
R205	3	994.42	[25]	3	1059.91	3.83	1029.55
	5	954.16	[26]	**4**	**964.02**		
R206	3	906.142	[25]	3	931.762	3	994.92
R207	2	890.61	[25]			3	896.72
	3	814.78	[33]	3	855.37		
R208	2	726.82	[25]	**3**	**708.9**	3	740.94
	4	698.88	[34]				
R209	3	909.16	[25]	3	983.75	3.93	920.18
	5	860.11	[26]	4	871.63		
R210	3	939.37	[25]	3	978.11	3.63	992.18
				4	**935.01**		
R211	2	885.71	[25]			3	828.81
	4	761.1	[31]	**3**	**794.04**		
RC101	14	1696.94	[25]			15.6	1652.38
	15	1619.8	[34]	15	1624.97		
RC102	12	1554.75	[25]			13.97	1497.056
	13	1470.26	[27]	13	1497.43		
	14	1466.84	[26]	14	1467.25		
RC103	11	1261.67	[25]	11	1265.86	11.8	1284.24

Table 5. *(Continued)*

Instance	Best Known			VNS-C			
	NV	TD	Ref.	Best		Average	
				NV	TD	NV	TD
RC104	10	1135.48	[25]	10	1136.49	10.7	1171.61
RC105	13	1629.44	[25]			15.6	1570.33
	14	1589.91	[27]	14	1642.81		
	15	1513.7	[26]	15	1524.14		
RC106	11	1424.73	[25]	**12**	**1396.59**	13.07	1408.7
	13	1371.69	[27]	13	1376.99		
RC107	11	1230.48	[25]	11	1254.68	11.93	1258.32
	12	1212.83	[26]	12	1233.58		
RC108	10	1139.82	[25]			11	1149.38
	11	1117.53	[26]	11	1131.23		
RC201	4	1406.94	[25]	4	1457.87	4.93	1365.76
	6	1134.91	[27]	5	**1310.44**		
RC202	3	1365.64	[25]			4	1278.96
	4	1181.99	[31]	4	1219.49		
RC203	3	1049.62	[25]			4	1020.716
	4	1026.61	[27]	4	**957.1**		
RC204	3	798.46	[25]	3	829.13	3	867.85
RC205	4	1297.65	[25]			5	1273.03
	5	1295.46	[27]	5	**1233.46**		
RC206	3	1146.32	[25]			4	1152.29
	4	1139.55	[27]	4	**1107.4**		
RC207	3	1061.14	[25]			4	1084.44
	4	1079.07	[33]	4	**1032.78**		
RC208	3	828.14	[25]	3	830.06	3	922.47

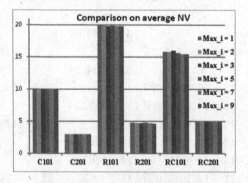

Fig. 3. Comparison on average NV with different Max_i values of $LinkMove\text{-}i$.

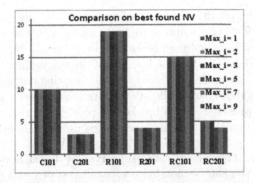

Fig. 4. Comparison on the best found NV with different Max_i values of $LinkMove\text{-}i$.

Table 6. Gaps of two algorithm variants with the VNS-C.

Instances		C101	C201	R101	R201	RC101	RC201
Gaps without	Average TD	0.00%	0.00%	0.46%	6.23%	3.62%	3.36%
CROSS-i	Best TD	0.00%	0.00%	0.02%	0.52%	2.00%	2.35%
Gaps without	Average TD	0.00%	0.00%	0.02%	2.07%	1.14%	0.63%
Or-opt-i	BestTD	0.00%	0.00%	0.00%	1.53%	0.79%	0.00%

the original VNS-C's results on TD. It can be found that, both algorithm variants' performance decreases in varying degrees on sample instances. Therefore, both operators are essential to VNS-C and no one is redundant. In addition, it can also be found that $CROSS\text{-}i$ plays a greater role than $Or\text{-}opt\text{-}i$ on optimizing the TD objective, since the generated solutions will show more obvious setbacks (larger gap) when $CROSS\text{-}i$ is removed.

4   Conclusions

Local search methods have not been widely used in addressing problems with multiple objectives due to their lack of efficiency in addressing multiple objectives simultaneously. To simultaneously optimize both objectives in VRPTW, a variable neighbourhood search (VNS) algorithm with compound neighbourhood operators (VNS-C) is developed in this paper. In the proposed algorithm, a new way of compounding the independent operators concerning both operation position and exchange link length is adopted in three compound neighbourhood operators. Two proposed compound neighbourhood operators are developed based on the $Or\text{-}opt$ and $CROSS$ exchange operators by considering specific features in VRPTW. Another neighbourhood operator $LinkMove\text{-}i$, which can reduce the number of vehicles simultaneously, is also proposed.

Comparing with the best known solutions in the current literature, experiment results on benchmark datasets show that VNS-C can produce promising results for

VRPTW. Especially, on the instances where customers are clustered or route lengths are long, VNS-C generates higher quality solutions. The long-route instances, i.e. with wide scheduling horizon or large vehicle capacity, can be efficiently addressed with VNS-C. Besides, VNS-C shows stronger performance in minimizing the total travel distance compared to reducing the number of vehicles. Two-phase methods showed to be effective in obtaining a lower number of required vehicles. Hybrid approaches, which invoke two-phase algorithms and other effective operators based on VNS, remain a promising direction in our future work.

Acknowledgement. This research was supported by Royal Society International Exchanges Scheme, National Natural Science Foundation of China (NSFC 71471092, NSFC-RS 713 11130142), Ningbo Sci&Tech Bureau (2014A35006), Department of Education Fujian Province (JB14223) and School of Computer Science at the University of Nottingham.

References

1. Laporte, G.: The vehicle routing problem: an overview of exact and approximate algorithms. Eur. J. Oper. Res. **59**, 345–358 (1992)
2. Cordeau, J.F., Desaulniers, G., Desrosiers, J., Solomon, M., Soumis, F.: VRP with time windows. In: The vehicle routing problem, Society for Industrial and Applied Mathematics, pp. 157–193 (2001)
3. Bräysy, O.: A reactive variable neighborhood search for the vehicle-routing problem with time windows. INFORMS J. Comput. **15**, 347–368 (2003)
4. Ghoseiri, K., Ghannadpour, S.F.: Multi-objective vehicle routing problem with time windows using goal programming and genetic algorithm. Appl. Soft Comput. **10**, 1096–1107 (2010)
5. Jozefowiez, N., Semet, F., Talbi, E.G.: Multi-objective vehicle routing problems. Eur. J. Oper. Res. **189**, 293–309 (2008)
6. Solomon, M.M.: Algorithms for the vehicle routing and scheduling problems with time window constraints. Oper. Res. **35**, 254–265 (1987)
7. Bräysy, O., Gendreau, M.: Metaheuristics for the vehicle routing problem with time windows. Report STF42 A 1025 (2001)
8. Potvin, J.Y., Kervahut, T., Garcia, B.L., Rousseau, J.M.: The vehicle routing problem with time windows part I: tabu search. INFORMS J. Comput. **8**, 158–164 (1996)
9. Van Breedam, A.: Improvement heuristics for the vehicle routing problem based on simulated annealing. Eur. J. Oper. Res. **86**, 480–490 (1995)
10. Hansen, P., Mladenović, N., Pérez, J.A.M.: Variable neighbourhood search: methods and applications. Ann. Oper. Res. **175**, 367–407 (2010)
11. Mladenović, N., Hansen, P.: Variable neighborhood search. Comput. Oper. Res. **24**, 1097–1100 (1997)
12. Hansen, P., Mladenović, N.: J-means: a new local search heuristic for minimum sum of squares clustering. Pattern Recogn. **34**, 405–413 (2001)
13. Hansen, P., Mladenović, N., Perez-Britos, D.: Variable neighborhood decomposition search. J. Heuristics **7**, 335–350 (2001)
14. Hansen, P., Mladenović, N., Urošević, D.: Variable neighborhood search and local branching. Comput. Oper. Res. **33**, 3034–3045 (2006)

15. Polacek, M., Hartl, R.F., Doerner, K., Reimann, M.: A variable neighborhood search for the multi depot vehicle routing problem with time windows. J. Heuristics **10**, 613–627 (2004)
16. Fleszar, K., Osman, I.H., Hindi, K.S.: A variable neighbourhood search algorithm for the open vehicle routing problem. Eur. J. Oper. Res. **195**, 803–809 (2009)
17. Hemmelmayr, V.C., Doerner, K.F., Hartl, R.F.: A variable neighborhood search heuristic for periodic routing problems. Eur. J. Oper. Res. **195**, 791–802 (2009)
18. Lin, S.: Computer solutions of the traveling salesman problem. Bell Syst. Tech. J. **44**, 2245–2269 (1965)
19. Or, I.: Traveling salesman-type combinatorial problems and their relation to the logistics of regional blood banking. Xerox University Microfilms (1976)
20. Osman, I.H.: Metastrategy simulated annealing and tabu search algorithms for the vehicle routing problem. Ann. Oper. Res. **41**, 421–451 (1993)
21. Taillard, A., Badeau, P., Gendreau, M., Guertin, F.A., Potvin, J.Y.: A tabu search heuristic for the vehicle routing problem with soft time windows. Transp. Sci. **31**, 170–186 (1997)
22. Ergun, Ö., Orlin, J.B., Steele-Feldman, A.: Creating very large scale neighborhoods out of smaller ones by compounding moves. J. Heuristics **12**, 115–140 (2006)
23. Rego, C.: A subpath ejection method for the vehicle routing problem. Manag. Sci. **44**, 1447–1459 (1998)
24. Dueck, G.: New optimization heuristics: the great deluge algorithm and the record-to-record travel. J. Comput. Phys. **104**, 86–92 (1993)
25. SINTEF: Best known solution values for solomon benchmark (2015). http://www.sintef.no/Projectweb/TOP/VRPTW/Solomon-benchmark/100-customers/
26. Alvarenga, G.B., Mateus, G.R., De Tomi, G.: A genetic and set partitioning two-phase approach for the vehicle routing problem with time windows. Comput. Oper. Res. **34**, 1561–1584 (2007)
27. Tan, K., Chew, Y., Lee, L.: A hybrid multiobjective evolutionary algorithm for solving vehicle routing problem with time windows. Comput. Optim. Appl. **34**, 115–151 (2006)
28. Küçükoğlu, İ., Öztürk, N.: An advanced hybrid meta-heuristic algorithm for the vehicle routing problem with backhauls and time windows. Comput. Ind. Eng. **86**, 60–68 (2014)
29. Kallehauge, B., Larsen, J., Madsen, O.B.: Lagrangian duality applied to the vehicle routing problem with time windows. Comput. Oper. Res. **33**, 1464–1487 (2006)
30. Cook, W., Rich, J.L.: A parallel cutting-plane algorithm for the vehicle routing problem with time windows. Computational and Applied Mathematics Department, Rice University, Houston, TX, Technical report (1999)
31. Ombuki, B., Ross, B.J., Hanshar, F.: Multi-objective genetic algorithms for vehicle routing problem with time windows. Appl. Intell. **24**, 17–30 (2006)
32. Chiang, W.C., Russell, R.A.: A reactive tabu search metaheuristic for the vehicle routing problem with time windows. INFORMS J. Comput. **9**, 417–430 (1997)
33. Rochat, Y., Taillard, É.D.: Probabilistic diversification and intensification in local search for vehicle routing. J. Heuristics **1**, 147–167 (1995)
34. Ursani, Z., Essam, D., Cornforth, D., Stocker, R.: Localized genetic algorithm for vehicle routing problem with time windows. Appl. Soft Comput. **11**, 5375–5390 (2011)

Product Modularization Using Cuckoo Search Algorithm

Hayam G. Wahdan, Sally S. Kassem,
and Hisham M.E. Abdelsalam[(⊠)]

Faculty of Computers and Information, Cairo University, Giza, Egypt
{hayam, s.kassem, h.abdelsalam}@fci-cu.edu.eg

Abstract. Modularity plays an important role in managing complex systems by dividing them into a set of modules that are interdependent within and independent across the modules. In response to the changing market trend of having large varieties within small production processes, modular design has assumed significant roles in the product development process. The product to be designed is represented in the form of a Design Structure Matrix (DSM). The DSM contains a list of all product components and the corresponding dependency patterns among these components. The main objective of this paper is to support design of products under modularity through clustering products into a set of modules or clusters. In this research, Cuckoo Search optimization algorithm is used to find the optimal number of clusters and the optimal assignment of components to clusters. The objective is minimizing the total coordination cost. Results obtained show an improved performance compared to published studies.

Keywords: Design structure matrix · Cuckoo search · Modularity · Modular design · Clustering · Optimization

1 Introduction

System design involves clustering various components in a product such that the resulting modules are effective for the company. An ideal architecture is one that partitions the product into practical and useful modules. Some successfully designed modules can be easily updated on regular time cycles, some can be made in multiple levels to offer wide market variety, some can be easily removed as they stay, and some can be easily swapped to gain added functionality. The importance of effective product modularity is multiplied when identical modules are used in various different products [1].

Modularization in product design can help speed up the new product development process [2]. The product is represented in the form of a Design Structure Matrix (DSM) that contains a list of all product components and the corresponding information exchange and dependency patterns.

DSM, working as a product representation tool, provides a clear visualization of product design. The transformation of Component-DSM into proposed functional blocks of components is called clustering. For small problems' components, a Component-DSM may be sorted manually. For larger problems, this is not practical, and at some point, computer algorithms are absolutely necessary [3].

© Springer International Publishing AG 2017
B. Vitoriano and G.H. Parlier (Eds.): ICORES 2016, CCIS 695, pp. 20–34, 2017.
DOI: 10.1007/978-3-319-53982-9_2

The aim of this paper is to develop a cuckoo search (CS) optimization algorithm to find: (1) the optimal number of clusters in a DSM; and (2) the optimal assignment of components to each cluster. The objective function is to minimize the total coordination cost. In this context, the DSM will work as a system analysis tool that provides a compact and clear representation of a complex system. It captures the interactions/interdependencies/interfaces between system elements. It also works as a project management tool which renders a project representation that allows for feedback and cyclic activity dependencies [4].

The rest of the paper is structured as follows. Section 2 provides a brief introduction on DSM. Section 3 reviews the literature and introduces the previous work this research builds on. Section 4 provides the problem definition. Section 5 presents the proposed algorithm. Section 6 presents and discusses the results obtained and, finally, Sect. 7 provides conclusion and ideas for future research.

2 Design Structure Matrix

The design structure matrix (DSM) is becoming a popular representation and analysis tool for system modeling, A DSM displays the relationships between components of a system or product in a compact visualization. Such a system can be, for example, product architecture or an engineering design process or a project.

The basic DSM is a simple square matrix, of size n, where n is the number of system elements. An example of a DSM is shown in Fig. 1. Element names are placed on the left hand side of the matrix as row headings and across the top row as column headings in the same order. If an element i depends on element j, then the matrix element ij (row$_i$, column$_j$) contains "1" or "x" otherwise the cell contains "0" or empty cell [5].

Once the DSM for a product is constructed, it can be analyzed for identifying modules, a process referred to as clustering. The goal of DSM clustering is to find a clustering arrangement where modules minimally interact with each other, while components within a module maximally interact with each other. As an example,

(a)   (b)

Fig. 1. Design structure matrix.

consider the DSM shown in Fig. 1(a). One can see from Fig. 1(b) that the DSM is rearranged by permuting rows and columns to contain most of the interactions within two separate modules: {A, F, E} and {D, B, C, G}. However, three interactions are left out of any modules.

3   Related Work

The idea of maximizing interactions within modules and minimizing interactions between modules within a DSM was proposed by [6]. A stochastic clustering algorithm using this principle operating on a DSM was first introduced in [7], with subsequent improvements presented by [8]. The proposed algorithm can find clustering solutions to architecture and organization interaction problems modeled using DSM method. Gutierrez (1998) developed a mathematical model to minimize the coordination cost, and hence, find the optimal solution for a given number of clusters. A Simulated annealing algorithm was performed by [9] to find clustered DSM with cost minimization as an objective.

Yassine et al. (2007) used the design structure matrix (DSM) to visualize the product architecture, and to develop the basic building blocks required for the identification of product modules. The clustering method was based on the minimum description length (MDL) principle and a simple genetic algorithm (GA) [10].

Borjesson (2009) proposed a method for promoting better output from the clustering algorithm used in the conceptual module generation phase by adding convergence properties, a collective reference to data identified as option properties, geometrical information, flow heuristics, and module driver compatibility [11].

Van Beek et al. (2010) developed a modularization scheme based on the functional model of a system. The k-means clustering was adopted for DSM based modularization by defining a proper entity representation, a relation measure and an objective function [12]. A novel clustering method utilizing Neural Network algorithms, and Design Structure Matrices (DSMs) was introduced by (Pandremenos and Chryssolouris 2012). The algorithm aimed to cluster components in DSM with predetermined number of clusters and clustering efficiency as an objective function [13].

Borjesson and Hölttä (2012) used IGTA (Idicula-Gutierrez-Thebeau Algorithm) for clustering Component-DSM as the basis for their work. They provided some improvement named IGTA-plus. IGTA-plus represented a significant improvement in the speed and quality of the solution obtained [3].

Borjesson and Sellgren (2013) presented an efficient and effective Genetic clustering algorithm, with the Minimum Description Length measure. To significantly reduce the time required for the algorithm to find good clusters, a knowledge aware heuristic element is included in the GA process. The efficiency and effectiveness of the algorithm is verified with four case studies [14].

Yang et al. (2014) provided a systematic clustering method for organizational DSM. The proposed clustering algorithm was able to evaluate the clustering structure based on the interaction strength [15].

Jung and Simpson (2014) introduced simple new metrics that can be used as modularity indices bounded between 0 and 1, and also utilized as the objective functions to obtain the optimal DSM. The optimum DSM was the one with the maximized interactions within modules and the minimized interactions between modules [16].

Kim et al. (2015) provided a new approach for product design by integrating assembly and disassembly sequence structure planning [17].

The literature on DSM design shows that there are a few techniques available to cluster DSM for modularity. The main difference among these techniques is the clustering objective. Cost minimization is one of the first clustering objectives, where each DSM element is placed in an individual cluster, then, components are coordinated across modules. The objective is to minimize the cost of being inside and outside a cluster. A clustered DSM can be compared to a targeted DSM topology using another objective function called Minimal Description Length (MDL). MDL finds mismatching elements between the two topologies. The objective of clustering is to minimize MDL. The number of clusters is determined a priori based on the DSM structure. Clustering Efficiency (CE) index is another clustering objective that evaluates a weighed count of zero elements inside clusters and non-zero elements outside clusters with a predefined number of clusters.

4 Problem Definition

The problem presented in this work considers two decision variables: (1) the number of clusters to be formed and (2) the optimal assignment of elements to each cluster. The objective function is to minimize the total coordination cost. The total coordination cost of the DSM to be clustered is based on IntraClusterCost and ExtraClusterCost as shown in Eqs. 1 and 2,

$$\text{IntraClusterCost} = \sum_{i,k \in \text{Cluster}j} (\text{DSM}_{ik} + \text{DSM}_{ki}) * \text{Clustersize}(j)^{\text{powcc}}, \quad (1)$$

$$\text{ExtrClusterCost} = \sum_{i,k \notin \text{cluster}j} (\text{DSM}_{ik} + \text{DSM}_{ki}) * \text{DSMSize}^{\text{powcc}}, \quad (2)$$

$$j = 1 \ldots \text{ncluster}$$

where DSM_{ik} is the coupling between elements i and k, DSMSize is the number of elements (rows) in the matrix, powcc is the exponent used to penalize the size of clusters, and ncluster is the total number of clusters. clustersize is the number of elements in cluster j [18].

$$\text{Total coordination Cost} = \text{ExtraClusterCost} + \text{IntraClusterCost}$$

Subject to the constraint that each element is assigned only to one cluster, in other words, overlap between clusters is not allowed. Prohibiting overlap, or multi-cluster elements, is important for the following reasons: when allowing elements to be assigned in multiple clusters, the importance and usefulness of the clustering algorithm will be diminished or eliminated. If elements exist in more than one cluster, this forces interactions between these clusters on multi levels. It is advantageous that elements placed in the same cluster are very similar [13].

Modularity affects both the profit and the sustainability of the product. A modular product contains modules that can be removed and replaced. The manufacturer can develop new modules instead of entirely new products. Therefore, customers buying upgraded modules only dispose of a portion of the product, thus reducing the total amount of waste. Hence, a customer upgrading a module does not have an entirely new product.

5   Proposed Algorithm

5.1   Cuckoo Search Algorithm and Pseudo Code

Yang and Deb (2009) proposed a new Meta heuristic algorithm called cuckoo search (CS). They tried to simulate the behavior of cuckoos to examine the solution space for optimization. The algorithm was inspired by the obligate interspecific brood parasitism of some cuckoo species that lay their eggs in the nests of host birds of other species. The aim is to escape the parental investment in raising their offspring. This strategy is also useful to minimize the risk of egg loss to other species, as the cuckoos can distribute their eggs amongst a number of different nests [19].

Of course, sometimes it happens that the host birds discover the alien eggs in their nests. In such case, the host bird takes different responsive actions varying from throwing such eggs away, to simply leaving the nest and building a new one elsewhere. On the other hand, the brood parasites have their own sophisticated characteristics to ensure that the host birds will care for the nestlings of their parasites. Examples of these characteristics are shorter egg incubation periods, rapid nestling growth, and egg coloration or pattern mimicking their hosts [20].

One major advantage of CS is its efficiency. The efficiency of the CS had been proven using many testing functions, for example, Michaelwicz function, Rosenbrock's function, etc. When comparing results with existing GA and PSO's, cuckoo search performed better [21]. Another major advantage of CS when compared to other metaheuristic algorithms, is its simplicity since it requires only two parameters. This feature reduces the effort of adjustment and fine tuning of parameter settings.

In cuckoo search, each egg can be regarded as a solution. In the initial process, each solution is generated randomly. When generating the ith solution in t + 1 generation, denoted by X_i^{t+1} a levy flight is performed as shown in Eq. 3,

$$X_i^{t+1} = X_i^t + \alpha \oplus \text{Levy}(\lambda) \tag{3}$$

Where $\alpha > 0$ is a real number denoting the step size, which is related to the sizes of the problem of interest, and the $\oplus$ product denotes entry-wise multiplications. A Levy flight is a random walk where the step-lengths are distributed according to a heavy-tailed probability distribution as shown in Eq. 4. The random walk via Lévy flight is more efficient in exploring the search space as its step length is much longer in the long run.

$$\text{Levy} \sim u = t - \lambda, \ (1 < \lambda \leq 3) \tag{4}$$

The CS algorithm is based on three idealized rules [22].

(1) Each cuckoo lays one egg at a time and dumps it in a randomly chosen nest.
(2) The best nests with high quality eggs (solutions) will be carried over to the next generations.
(3) The number of available host nests is fixed, and a host can discover an alien egg with a probability $pa \in [0, 1]$. In this case, the host bird can either throw the egg away or abandon the nest to build a completely new nest in a new location.

For simplicity, the third assumption can be approximated by a fraction pa of the n nests being replaced by new nests (with new random solutions at new locations). For a maximization problem, the quality or fitness of a solution could be proportional to the objective function. However, other more sophisticated expressions for the fitness function can also be defined.

Based on these three rules, the basic steps of the CS algorithm are summarized in the pseudo code in Fig. 2.

5.2 Solution Representation

The CS algorithm is used to solve the problem defined in Sect. 4. Solution representation of the problem is a vector of size equals to the number of elements in the DSM. Each cell in the vector takes an integer value between 1 and the number of elements, as

---

Objective function f(x), x = (x1, ..., xd)^T;

Initial population of n host nests xi (i = 1, 2, ..., n);

while (t <MaxGeneration) or (stop criterion);

Get a cuckoo (i) randomly using Levy flights;

Evaluate its quality/fitness Fi;

Choose a nest among n (j) randomly;

if (Fi > Fj),

Replace j with the new solution;

end

Abandon a fraction (pa) of worse nests

and build new ones at new locations via Levy flights;

Keep the best solutions (or nests with quality solutions);

Rank the solutions and find the current best;

end while

Postprocess results and visualisation;

Fig. 2. Pseudocodeof CS [23].

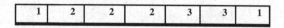

1	2	2	2	3	3	1

Fig. 3. Solution representation vector.

show in Fig. 3. The vector in Fig. 3 with size 7 represents a solution, where the DSM Contains 7 elements. Elements 1 and 7 belong to cluster 1, elements 2, 3, 4 belong to cluster 2, and elements 5, 6 belong to cluster 3. This solution representation forces the element to be a member of only one cluster.

Assume that we start with the maximum possible number of clusters, which equals to the number of elements in the DSM. The next step is to try to find the optimal number of clusters after deleting empty clusters. Such representation of the problem will not allow multi-clustering, which means each element will be assigned to only one cluster.

The problem is solved using Cuckoo search algorithm (CS). CS solves continuous types of variables. Since the problem in hand is categorized as a discrete variable problem, the solutions should be converted from continuous to discrete. This is done by the discretization of the continuous space by transforming the values into a limited number of possible states. There are several discretization methods available in the literature, for example:random key technique is used to transform from continues space to discrete integer space. In order to decode the position, the nodes are visited in ascending order for each dimension [24]. Another discretization methods is the smallest position value (SPV) method. The technique maps the positions of the solution vector by placing the index of the lowest valued component as the first item on a permutated solution. The second lowest value component is the second item, and so on [25]. The nearest integer method is another technique, to transform continuous variables to integer variables. In the nearest integer method, a real value is converted to the nearest integer (NI) by rounding or truncating up or down [26].

Considering the above mentioned methods, SPV, and random key methods, are not suitable for the problem presented in this work. This is because integer value(s) need to be repeated, while these methods result in unique values. Therefore, the suitable method for the problem in hand is the nearest integer method since it allows the repetition of values by truncating to the higher or lower value.

To cluster a DSM into modules using CS, we start with a set of nests; each nest is a vector of length that equals to the number of elements in the DSM. This vector contains random numbers following uniform distribution in the range from lower and upper limits. These random numbers are converted to integer values using the nearest integer method. Each one of these integer numbers represent a solution that could be sent for the evaluation function. The evaluation function returns the total coordination cost.

5.3   Solution Evaluation

The total coordination cost of the DSM to be clustered is based on IntraClusterCost and ExtraClusterCost as explained in Sect. 4. Regarding intracluster cost, if interaction DSM_{ik} belongs to cluster j, then calculate intra cluster cost. On the other hand, if interaction DSM_{ik} does not belong to cluster j, calculate the extra cluster cost. The first step in calculating the total coordination cost is to start with the total number of interactions in the DSM multiplied by the size of the DSM raised to the power powcc. This is the highest value of total coordination. This value will be minimized in subsequent steps of the algorithm after forming clusters. After completion of the evaluation step, select the best solution and go to the next best solution using Levy flight, carrying the best nests with high quality eggs (solutions) over to the next generations. Continue till the stopping condition is reached.

6 Experimental Results and Analysis

In this section we examine the CS algorithm on a number of instances. The CS algorithm has two parameters namely, the (pa) value. The (pa) value represents the probability of discovering an alien egg, which corresponds to getting rid of solution and the number of nests corresponds to the population of solutions. There are no general recommendations in the literature on the range(s) of values for these two parameters. Therefore, we examine the solution quality of the test instances over a range of possible values for the two parameters. We chose the different values of (pa) in the range [0.1, 0.9], with an increment of 0.1. We also examine the solution quality with different number of nests, namely, 25, 50, 75, and 100. We noticed that the change in the number of nests did not have a significant impact on the objective function value in all tested cases. The maximum number of iterations is 3000 and it is used as the stopping condition. The algorithm is coded using Matlab, and is run on a computer withIntel(R) Core(TM) i3 CPU, 2.27 GHz PC. We use three test instances, 2 are categorized as small size instances with 7 and 9 elements each, available in [9, 10], respectively. The third instance is larger in size and consists of 61 elements, and is available in [9].

The first small size instance has a DSM that contains 7 elements as shown in Fig. 4. The DSM starts initially with a total coordination cost of 68. This cost is based on assigning each element in it is own cluster. No clusters are formed yet.

After applying the CS clustering algorithm, the clustered DSM is as shown in Fig. 5. The Total coordination cost is reduced to 48. This problem was solved in [9] and a total coordination cost of 53 was obtained. Hence, the proposed CS is able to obtain better results. The minimum number of clusters using the proposed CS algorithm is 2. Figure 6 shows the total cost as it changes with every iteration. The best solution is obtained in iteration number 575. The CPU run time ranges from 0.07 s to 0.66 s for 100 to 3000 iterations, respectively.

It is noticed that, in the clustered DSM two clusters are formed and the majority of the interactions are included in clusters. This indicates that similar elements are grouped in the same cluster. In this case, IntraClusterCost is larger than the ExtraClusterCost which improves the objective function value. Only 3 interactions are placed outside clusters (number of 1's).

Figure 7 shows the objective function values corresponding to different (pa) values. It is noticed from Fig. 7 that the objective function value remains 48 till the value of (pa) reaches 0.5. When the value of (pa) exceeds 0.5, the objective function value

	1	2	3	4	5	6	7
1		1	0	0	1	1	0
2	0		0	1	0	0	1
3	0	1		1	0	0	1
4	0	1	1		1	0	1
5	0	0	0	1		1	0
6	1	0	0	0	1		0
7	0	1	1	1	0	0	

Fig. 4. Original DSM.

	1	5	6	2	3	4	7
1		1	1	1	0	0	0
5	0		1	0	0	1	0
6	1	1		0	0	0	0
2	0	0	0		0	1	1
3	0	0	0	1		1	1
4	0	1	0	1	1		1
7	0	0	0	1	1	1	

Fig. 5. Clustered DSM.

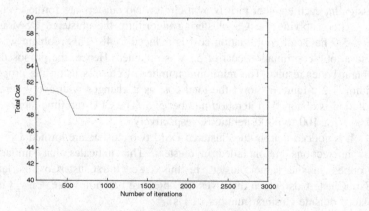

Fig. 6. Cost history for CS algorithm-best solution.

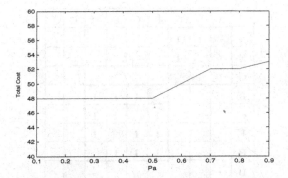

Fig. 7. The relation between Pa and total cost.

deteriorates. This is due to the fact that high values of (pa) tend to get rid of solutions without trying to improve them locally. Therefore, the value of (pa) in the range [0.1, 0.5] for this test instance, achieves the required balance between exploration and exploitation.

We examined the developed algorithm on another problem presented in [10]. The DSM of the problem has 9 elements as shown in Fig. 8.

Figure 9 shows the clustered DSM after using CS algorithm. The total coordination cost after clustering with CS is 41.8. The corresponding number of clusters is 4. The resulting DSM clustered using our proposed CS algorithm is the same as the one obtained in [10]. Figure 10 shows the total cost as it changes with every iteration. The best solution is obtained in iteration number 880. The CPU run time ranges from 1.07 s to 13.52 s for 100 to 3000 iterations, respectively.

We notice from Fig. 9 that, in the clustered DSM four clusters are formed, cluster 1 with the most similar 3 elements, cluster 2 with the most similar 4 elements, cluster 3 with 1 element and cluster 4 with 1 element. All interactions are included in clusters. In this case, there are no extra costs because no 1's are outside clusters.

	A	B	C	D	E	F	G	H	I
A	1	0	0	0	1	0	1	0	0
B	0	1	1	0	0	1	0	1	0
C	0	1	1	0	0	1	0	1	0
D	0	0	0	1	0	0	0	0	0
E	1	0	0	0	1	0	1	0	0
F	0	1	1	0	0	1	0	1	0
G	1	0	0	0	1	0	1	0	0
H	0	1	1	0	0	1	0	1	0
I	0	0	0	0	0	0	0	0	1

Fig. 8. Original DSM.

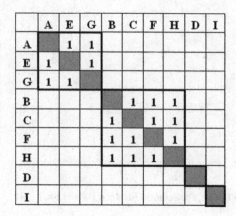

	A	E	G	B	C	F	H	D	I
A		1	1						
E	1		1						
G	1	1							
B					1	1	1		
C				1		1	1		
F				1	1		1		
H				1	1	1			
D									
I									

Fig. 9. Clustered DSM.

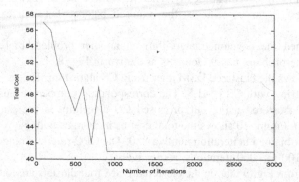

Fig. 10. Cost history for CS algorithm-best solution.

Figure 11 shows the objective function values corresponding to different (pa) values. It is noticed from Fig. 11 that the objective function value remains 41.8 till the value of (pa) reaches 0.6. When the value of (pa) exceeds 0.6, the objective function value deteriorates. The impact of the value of (pa) on solution quality is similar to that in the first test instance.

To further evaluate the proposed CS algorithm we apply it on a large size problem, available in [9]. The DSM contains 61 elements and represents an elevator example. The total coordination cost obtained using the CS algorithm is 4133.25, with a total number of 17 clusters. The total coordination cost obtained in [9] is 4433. Accordingly, our proposed CS algorithm is able to obtain superior results when compared to the results obtained by [9]. Cluster assignments of the elevator example using the CS algorithm are shown in Table 1. The best solution is obtained in iteration number 921. The CPU run time is 450.7 s after 3000 iterations. Figure 12 shows the total cost as it changes with every iteration.

Table 1 shows that 17 clusters are formed, each cluster contains the most similar elements which minimizes the total coordination cost. Figure 13 shows the impact of

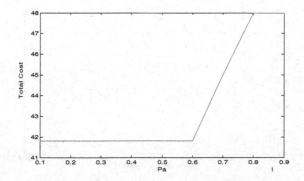

Fig. 11. The relation between Pa and total cost.

changing the values of (pa) on the objective function value. Similar to the previous test instances, it is noticed that the solution deteriorates as the value of (pa) exceeds a certain value. For this test instance the (pa) value after which the solution starts to deteriorate is 0.4.

Table 1. Results obtained using CS algorithm for the elevator example.

Cluster number	Elements that cluster contains
1	1, 3, 5, 11, 15, 17, 18, 20, 22, 28, 34, 35, 37, 39, 40, 41, 43, 47, 48, 50, 59, 60, 61
2	2, 8, 12, 16, 19, 21, 26, 27, 32, 33, 44, 46, 49, 54
3	4
4	6, 9, 13
5	7
6	10, 14, 25, 55
7	20
8	22, 53
9	23, 31
10	24
11	30
12	36
13	51
14	52
15	56
16	57
17	58

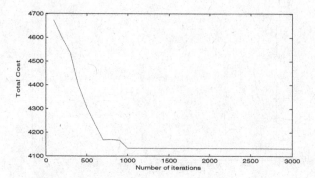

Fig. 12. Cost history for CS algorithm–best solution.

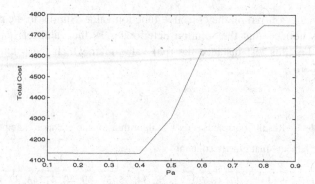

Fig. 13. The relation between Pa and total cost.

7   Conclusion and Future Work

The importance of modular product design has been recognized due to many reasons, for example, its ability to achieve economies of scale, providing high degree of flexibility during production, and reducing lead times. Clustering of a design structure matrix (DSM) is one of the methods that are widely used to design a product under modularity. In this work, we aimed at designing modular products through representation and clustering of their DSM. The clustering process required solving for 2 decision variables: the number of clusters and the elements assigned to each cluster. The objective function was minimizing the total coordination cost, under the constraint that each element was to be assigned to one cluster, without allowing clusters' overlap. Heuristics and metaheuristic techniques are typically used to cluster DSMs for modularity. Cuckoo search (CS) is a metaheuristic algorithm that has proven efficiency in terms of solution quality, speed, and simplicity, when compared to other metaheuristics algorithms available in the literature. In this work, we adopted CS algorithm to perform DSM clustering. To test its performance, we applied CS on a number of instances available in the literature. The proposed CS algorithm was able to obtain better or similar results for all the test instances. Future work includes restricting the number of elements belonging to each cluster, considering inventory and supply chain decisions in the modular design process, and including the number of clusters in the objective function.

References

1. Aguwa, C.C., Monplaisir, L., Sylajakumar, P.A.: Effect of rating modification on a fuzzy-based modular architecture for medical device design and development. In: Advances in Fuzzy Systems (2012)
2. Gwangwava, N., Nyadongo, S., Mathe, C., Mpof, K.: Modular clusterization product design support system. Intl. J. Adv. Comput. Sci. Technol. (IJACST) 2(11), 8–13 (2013)
3. Borjesson, F., Hölttä-Otto, K.: Improved clustering algorithm for design structure matrix. In: ASME 2012 International Design Engineering Technical Conferences & Computers and Information in Engineering Conference, Chicago, IL, USA: IDETC/CIE 2012, pp. 1–10 (2012)
4. Abdelsalam, H.M., Rasmy, M.H., Mohamed, H.G.: A simulation-based time reduction approach for resource constrained design structure matrix. Intl. J. Model. Optimization 4(1), 51–55 (2014)
5. Abdelsalam, H., Bao, H.: A simulation-based optimization framework for product development cycle time reduction. IEEE Trans. Eng. Manage. 53(1), 69–85 (2006)
6. Eppinger, S., Whitney, D., Smith, R., Gebala, D.: A model based method for organizing tasks in product development. Res. Eng. Des. 6, 1–13 (1994)
7. Idicula, J.: Planning for Concurrent Engineering. Gintic Institute Research, Singapore (1995)
8. Gutierrez, C.I.: Integration analysis of product architecture to support effective team co-location. Masters thesis, Massachusetts Institute of Technology, Cambridge (1998)
9. Thebeau, R.E.: Knowledge Management of System Interfaces and Interactions for Product Development Process. Massachusetts Institute of Technology, System Design and Management Program (2001)
10. Yassine, A.A., Yu, T.-L., Goldberg, D.E.: An information theoretic method for developing modular architectures using genetic algorithms. Res. Eng. Design 18, 91–109 (2007)
11. Borjesson, F.: Improved output in modular function deployment using heuristics. In: International conference on Engineering Design, Stanford, USA, pp. 24–27 (2009)
12. van Beek, T.J., Erden, M.S., Tomiyama, T.: Modular design of mechatronic systems with function modeling. Mechatronics 20(8), 850–863 (2010)
13. Pandremenos, J., Chryssolouris, G.: A neural network approach for the development of modular product architectures. Intl. J. Comput. Integr. Manuf. 24, 1–8 (2012)
14. Borjesson, F., Sellgren, U.: fast hybrid genetic clustering algorithm for design structure matrix. In: 25th International Conference on Design Theory and Methodology. ASME, Portland (2013)
15. Yang, Q., Yao, T., Lu, T., Zhang, B.: An overlapping-based design structure matrix for measuring interaction strength and clustering analysis in product development project. IEEE Trans. Eng. Manage. 61(1), 159–170 (2014)
16. Jung, S., Simpson, T.W.: A clustering method using new modularity indices and genetic algorithm with extended chromosomes. In: DSM 14 Proceedings of the 16th International DSM conference: Risk and Change management in complex systems, pp. 167–176 (2014)

17. Kim, S., Baek, J.W., Moon, S.K., Jeon, S.M.: A new approach for product design by integrating assembly and disassembly sequence structure planning. In: Handa, H., Ishibuchi, H., Ong, Y.-S., Tan, K.C. (eds.). PALO, vol. 1, pp. 247–257. Springer, Heidelberg (2015). doi:10.1007/978-3-319-13359-1_20
18. Borjesson, F., Itta-Otto, K.H.: A module generation algorithm for product architecture based on component interactions and strategic drivers. Res. Eng. Des. **25**(1), 31–51 (2014)
19. Yang, X., Deb, S.: Cuckoo search via Levy flights. In: The World Congress on Nature and Biologically Inspired Computing (NABIC 2009), pp. 210–214. IEEE, Coimbatore (2009)
20. Li, X., Yin, M.: Modified cuckoo search algorithm with self adaptive parameter method. Inf. Sci. **298**, 80–97 (2015)
21. Yang, X.-S., Deb, S.: Engineering optimisation by cuckoo search. Intl. J. Math Model Numerical Optimization **1**(4), 330–343 (2010)
22. Navimipour, N.J., Milani, F.S.: Task scheduling in the cloud computing based on the cuckoo search algorithm. Intl. J. Model. Optimization **5**(1), 44–47 (2015)
23. Yildiz, A.R.: Cuckoo search algorithm for the selection of optimal machining parameters in milling operations. Int. J. Adv. Manuf. Technol. **64**(1), 55–61 (2013)
24. Chen, H., Li, S., Tang, Z.: Hybrid gravitational search algorithm with random-key encoding scheme combined with simulated annealing. Intl. J. Comput. Sci. Mob. Comput. **11**(6), 208–217 (2011)
25. Verma, R., Kumar, S.: DNA sequence assembly using continuous particle swarm optimization with smallest position value rule. In: First International Conference on Recent Advances in Information Technology, pp. 410–415 (2012)
26. Burnwal, S., Deb, S.: Scheduling optimization of flexible manufacturing system using cuckoo search-based approach. Intl. J. Adv. Manuf. Technol. **64**, 1–9 (2012)

A Family of Models for Finite Sequential Games Without a Predetermined Order of Turns

Rubén Becerril-Borja[✉] and Raúl Montes-de-Oca

Departamento de Matemáticas, Universidad Autónoma Metropolitana-Iztapalapa,
Av. San Rafael Atlixco 186, Col. Vicentina, 09340 Mexico City, Mexico
ruben.becerril@gmail.com, momr@xanum.uam.mx

Abstract. Situations where individuals interact for a long time in different periods of time are usually modelled with simultaneous games repeated for each period of interaction. Therefore, decisions are made at the same time. Even models that are sequential assume the knowledge of the order in which players make their decisions. In this work a basic model for games with finite strategies sets is presented where the order of turns is not known beforehand, but is represented by a random variable. Other models are derived from the first one, allowing the study of more general cases, by changing the assumptions on different components of the game. A variation for bayesian games is presented as well. For all these models a series of results is obtained which guarantees the existence of Nash equilibria. A theoretical example and a possible application for drafting athletes are shown as well.

Keywords: Sequential games · Stochastic games · Existence of Nash equilibria · Turn selection process · Bayesian games · Finite strategy sets

1 Introduction

Game theory is focused on studying situations in which several subjects make decisions that affect all of them, usually in the form of a utility reward. Essentially, any interaction that can be roughly described as the one above can be studied as a game.

In the classical theory (see [6,7,13]), the situations that can be analysed are deterministic in their rules, which means that a particular structure in which individuals will interact is established well before the players actually interact, and such structure has no random elements once the players are making decisions. That, in itself, provides a frame to study the game mathematically, and therefore it is possible to come up with the decisions that have to be taken in order to maximise the utility of each player, subject to the fact that each player can only make his own decisions and cannot rely on other players.

Another important characteristic to take into account to study a situation is the time frame it is set on. When decisions are made in discrete time, games can be divided in two sets: if players make their decisions at the same

© Springer International Publishing AG 2017
B. Vitoriano and G.H. Parlier (Eds.): ICORES 2016, CCIS 695, pp. 35–51, 2017.
DOI: 10.1007/978-3-319-53982-9_3

time, or at different times, but they are never aware of the selections made by the other players, it is a simultaneous game; if there is a certain order to make decisions imposed before the game starts, such that players in the future may know the choices previously made by others in order to adapt their behaviour, then it is a sequential game.

The first approach to introduce random elements was made in sequential games, where a player called Nature makes its decision before any other player made theirs. Because of the structure of sequential games, one could decide whether players would be aware of the behaviour of the Nature player, which allows to study situations with random externalities.

Nevertheless, once the game is afoot, there are no more random elements. This led to the study of stochastic games, which allowed the possibility of random events occurring between decisions taken in each period of time, as in [9]. To do this, the theory is focused on the study of simultaneous games that are repeated in each period of time, but which may move to different stages depending on a distribution that observes the action taken by each player and the current stage of the game. Since it is a sequence of simultaneous games, in each iteration every player gets a utility, and at the end, the utility of each player is calculated as the discounted sum of all the utilities.

In recent years, this approach has been expanded, for example, by studying games where for every time period there is a generation of players, who engage in a noncooperative game, but for every player in each game, there is a descendant that plays in the next iteration, creating a sort of family throughout time, which has to play cooperatively while dealing with their contemporaries, as can be seen in [2,3,11,14]. Another line of study has been the modelling of altruism as a means of obtaining utility in a different way, by taking into account that the decision also influences the future, or understanding altruism as something that benefits the future generations that play the game. This approach is studied in [10,12]. Other advances deal with the change of stages, and its effect on some of the characteristics of the game, for example, by restricting the options available to each player through the use of penal codes, that is, by making players themselves punish each other according to the deviations incurred by some of them, which is developed in [8]. All of these models are analysed within the context of repeated simultaneous games, and they compute the utilities as a discounted sum of winning obtained in each repetition of the game.

The work presented here is an effort to advance the study of certain situations in a frame of work that is sequential, where players take decisions in order, but where the order in which such decisions are made is not given from the beginning, but rather decided at each period of the game. The decision of whose turn it is made by a selection process, which, as perceived by each player, is random, so players can assign a probability distribution to model it. Each individual only receives a utility at the end of the game, which depends on the decisions made by all of the players. Once this model has been established, the existence of an adequate equilibrium will be proven, and the result will be extended to cover other models. The first one allows modifications for each player regarding the way in which the selection process is modelled, perhaps as they learn new information through the course of the game. The second one conditions the selection process to the decisions made by all players at all the previous turns. Meanwhile, in the third model the strategy sets are conditioned on the decisions made in previous turns. As far as we know, the models and the corresponding results presented here are a new approach to the theory of sequential games.

The structure of this work is as follows: Sect. 2 defines the models in the family of sequential games with turn selection process. The models in Subsects. 2.2 and 2.3 consider a fixed number of decisions made by each player during the game. In Subsect. 2.4 the base model is described and refined to obtain the models described in Subsects. 2.5–2.7. In Subsect. 2.8 a bayesian model is considered.

In Sect. 3, using the model of Subsect. 2.3 as a base, the results that guarantee the existence of Nash equilibria are proved. Those propositions and their proofs can be easily modified to be adapted for the other models presented in Sect. 2. These results have been adapted and improved from those presented in [4].

In Sect. 4 we present two examples of games that can be studied with these ideas. Subsection 4.1 presents a very simple game played between two players, whereas in Subsect. 4.2 we present a possible application in the context of a draft of athletes for sports leagues.

2 Sequential Games and Turn Selection Process

2.1 Notation and Terminology

Several concepts used in this article are standard in game theory and can be consulted in a wide selection of books, such as [7,13]. Subscripts in elements denote the player attributed to the element in question, superscripts denote the period of time considered, and subscripts in brackets are used for indexing sequences.

The main components of this model are:

- A set of **players** $\mathcal{N} = \{1, 2, \ldots, N\}$, with $N \in \mathbb{N}$ fixed.
- For each player j, $\mathcal{S}_j$ is a finite set of **pure strategies**, which contains all the possible decisions that can be made throughout the game. The set of all pure strategies in the game is denoted by $\mathcal{S} = \cup_{j=1}^{N} \mathcal{S}_j$.
- A fixed $T \in \mathbb{N}$ called the **horizon** of the game, which is the number of turns to be played.
- A **utility function** $u_j \colon \Sigma \to \mathbb{R}$ for each player j, where Σ is the cartesian product of $\mathcal{S}$ with itself T times.
- A probability density $p_j \colon \mathcal{N} \to [0, 1]$ of the distribution that models the **turn selection process** according to each player j.

Furthermore, $\mathcal{M}_j$ will be defined as the set of **mixed strategies** for every player j obtained from the set of pure strategies $\mathcal{S}_j$.

The first thing to observe in any of the following models is that, since no player has the knowledge of the turns at which they'll be making decisions, a strategy plan for each of them must consider an action for every turn. Given that all players can see the decisions taken in previous periods, the plan of each player at turn ℓ should be conditioned by a **scenario** $s^\ell = (s^{\ell-1}, \dots, s^1)$, which is made of all the decisions taken up to period $\ell - 1$. This way each player j has a **plan of conditioned strategies**, denoted by s_j that gives a strategy to follow for each period ℓ and for each possible scenario he encounters at period ℓ. The set of plans of conditioned strategies is denoted by $\mathcal{P}_j$ if only pure strategies are considered. If mixed strategies are allowed, then the sets are denoted by $\mathcal{Q}_j$ for each player j.

Finally, taking all the plans for each player, it is possible to build a vector of size N, with the jth component being a plan of strategies for player j. These vectors are the **profiles of conditioned strategies** of the game. The set of profiles made only of pure strategies plans is denoted by $\mathcal{P}$, whereas the set of profiles that allow mixed strategies plans is denoted by $\mathcal{Q}$.

In order to find a solution to the models, it is necessary to evaluate the quality of each profile for each player j, while allowing randomization for the turn selection process. Therefore, an expectation operator E_j is defined for each player j and each profile $x \in \mathcal{Q}$ by building the product measure of the probability distribution of the turn selection process and the plan of mixed strategies that is being evaluated.

2.2 One Turn per Player

In this model exactly one decision will be allowed per player, but taking into consideration that the period of time in which said decision is to be made is not known beforehand. For example, we could picture a written price auction, that is an auction where players approach exactly once a bidding sheet, to write their only offer for the item. The player that bids the largest amount wins the auctioned item.

To evaluate a certain profile $x = (x_1, \dots, x_n) \in \mathcal{Q}$, we define the expected utility of player j in x as

$$E_j(x) = \sum_{(n^1,\dots,n^N)\in\mathsf{P}(\mathcal{N})} \sum_{s^1\in\mathcal{S}_{n^1}} \cdots \sum_{s^N\in\mathcal{S}_{n^N}} u_j(s^N, \dots, s^1)$$

$$\times x_{n^N}(s^N \mid s^{N-1}, \dots, s^1)p_j(n^N)\cdots x_{n^1}(s^1)p_j(n^1), \tag{1}$$

where $\mathsf{P}(\mathcal{N})$ is the set of permutations of $\mathcal{N}$:

Based on this, a **Nash equilibrium** can be defined as a profile of strategies $x^* \in \mathcal{Q}$ such that, for every player $j \in \mathcal{N}$,

$$E_j(x^*) \geq E_j(x^*_{-j}, x_j) \tag{2}$$

for all $x_j \in \mathcal{Q}_j$, where x_{-j} is the partial profile that considers the plans of strategies of all players except j. This way, a profile (y_{-j}, z_j) means the plans of strategies of y are considered for every player but j, for whom the plan of strategies z_j is used. The definition of Nash equilibrium given by (2) will be used for all the models that follow, where, respectively, the expected utility function presented in each section is used.

2.3   Fixed Number of Turns per Player

The previous model will be generalized by allowing each player to make more than one decision. The number of decisions made, however, is fixed beforehand for each player, but as before, the periods in which these decisions are made are not known a priori. This can be exemplified by a written price auction with m_j bids allowed to player j, that is, player j can approach the bidding sheet at most m_j times to write m_j offers. The winner of the item is the player that bids the highest price.

If the number of decisions made throughout the game for each player is the corresponding component of the vector $\mathbf{m} = (m_1, \ldots, m_N)$, then the expected utility for player j when the profile $x = (x_1, \ldots, x_N) \in \mathcal{Q}$ is followed can be defined as

$$
E_j(x) = \sum_{(n^1,\ldots,n^T)\in\mathsf{P_m}(\mathcal{N})} \sum_{s^1\in\mathcal{S}_{n^1}} \cdots \sum_{s^T\in\mathcal{S}_{n^T}} u_j(s^T,\ldots,s^1)
$$

$$
\times\, x_{n^T}(s^T \mid s^{T-1},\ldots,s^1)p_j(n^T)\cdots x_{n^1}(s^1)p_j(n^1),
$$

(3)

where $\mathsf{P_m}(\mathcal{N})$ is the set of permutations of $\mathcal{N}$ with m_k repetitions of k, and $T = \sum_{j\in\mathcal{N}} m_j$.

2.4   Unknown Number of Turns per Player: Base Model

Now we shift our attention to the main model. Instead of having information of how many decisions each player will be making in the game, that information is hidden, and all players are allowed to potentially make as many decisions as the turn selection process permits them during the T periods of time of the game. We could see this as a written price auction, where after T bids have been placed, the item is given to the player with the highest bid. In the previous model it doesn't mean necessarily the last player that bid, since after that, lower offers may have been placed. Another example is a game of chess in T moves, where the selection of the player that will make the following move is made by the flip of a coin before each move is made. Imagine that if the flip is heads, then white moves, and if the flip is tails, then black moves. This allows for successive moves to be made by the same colour.

In this case, the expected utility function for player j, when $x \in \mathcal{Q}$ is the profile considered, is given by

$$
E_j(x) = \sum_{n^1\in\mathcal{N}} \sum_{s^1\in\mathcal{S}_{n^1}} \cdots \sum_{n^T\in\mathcal{N}} \sum_{s^T\in\mathcal{S}_{n^T}} u_j(s^T,\ldots,s^1)
$$

$$
\times\, x_{n^T}(s^T \mid s^{T-1},\ldots,s^1)p_j(n^T)\cdots x_{n^1}(s^1)p_j(n^1).
$$

(4)

2.5   Updated Models of the Turn Selection Process in Each Period

In the base model of Subsect. 2.3, each player was thought to be making his predictions of the behaviour of the turn selection process from the beginning of the game, and not changing thereafter. However, this a priori distribution may not be accurate throughout the game, or the player may learn new information of its behaviour by observing how players are selected at each turn. Therefore, a player has to be allowed to have different distributions for each period.

Instead of having a fixed probability density p_j, each player has a vector $(p_j^1, p_j^2, \ldots, p_j^T)$ of probability densities for each period. The expected utility for player j when confronted with profile x will be defined as

$$E_j(x) = \sum_{n^1 \in \mathcal{N}} \sum_{s^1 \in \mathcal{S}_{n^1}} \cdots \sum_{n^T \in \mathcal{N}} \sum_{s^T \in \mathcal{S}_{n^T}} u_j(s^T, \ldots, s^1)$$

$$\times x_{n^T}(s^T \mid s^{T-1}, \ldots, s^1) p_j^T(n^T) \cdots x_{n^1}(s^1) p_j^1(n^1). \tag{5}$$

2.6   Conditioned Turn Selections on Previous Decisions

In the base model, only the mixed strategies are dependent on the decisions made in the previous turns. To approach the usual modelling of games by stages every element should be allowed to change according to the decisions made previously. Though originally it was expected to be conditioned as a Markov-like structure, that is, the structure would depend only on the immediate previous decision made in the game, it is possible to generalize it, conditioning the turn selection process on every single decision taken so far. This also accounts for the case in which the process is conditioned only on its own behaviour, that is, on the selections made, not on the decisions taken by each player. Here it is possible to have the chess game in T turns, but where to reduce the likelihood of the appearance of large sequences of moves made by the same colour, the "coin" could change its probabilities, so the choosing of a player also reduces his probability of being chosen in the next period.

In this case the expected utility of player j to evaluate a profile $x \in \mathcal{Q}$ is given by

$$E_j(x) = \sum_{n^1 \in \mathcal{N}} \sum_{s^1 \in \mathcal{S}_{n^1}} \cdots \sum_{n^T \in \mathcal{N}} \sum_{s^T \in \mathcal{S}_{n^T}} u_j(s^T, \ldots, s^1)$$

$$\times x_{n^T}(s^T \mid s^{T-1}, \ldots, s^1) p_j(n^T \mid s^{T-1}, \ldots, s^1) \cdots x_{n^1}(s^1) p_j(n^1). \tag{6}$$

2.7   Strategy Sets Changing According to Period

To approach the modelling per stages from a different perspective, the sets of strategies will be able to change according to the period of time in which the decision is made. This can also be modified to take into account the fact that strategy sets may change according to the previously made decisions.

For this model, the expected utility of player j for the profile $x \in \mathcal{Q}$ is defined as

$$
E_j(x) = \sum_{n^1 \in \mathcal{N}} \sum_{s^1 \in \mathcal{S}^1_{n^1}} \cdots \sum_{n^T \in \mathcal{N}} \sum_{s^T \in \mathcal{S}^T_{n^T}} u_j(s^T, \ldots, s^1)
$$

$$
\times x_{n^T}(s^T \mid s^{T-1}, \ldots, s^1) p_j(n^T) \cdots x_{n^1}(s^1) p_j(n^1)
$$

(7)

or, accordingly, the strategy sets are conditioned by the previously made decisions, that is:

$$
E_j(x) = \sum_{n^1 \in \mathcal{N}} \sum_{s^1 \in \mathcal{S}_{n^1}} \cdots \sum_{n^T \in \mathcal{N}} \sum_{s^T \in \mathcal{S}_{n^T}(s^{T-1}, \ldots, s^1)} u_j(s^T, \ldots, s^1)
$$

$$
\times x_{n^T}(s^T \mid s^{T-1}, \ldots, s^1) p_j(n^T) \cdots x_{n^1}(s^1) p_j(n^1).
$$

(8)

2.8 A Bayesian Model

In this model, there is a set of types Θ_j for each player j. From these, Nature chooses a type $\theta_j \in \Theta_j$. After each player knows his own type, each one has a distribution a priori $b_j(\cdot \mid \theta_j) \colon \Theta_{-j} \to [0, 1]$ for the types of the other players, for each $\theta_j \in \Theta_j$. The only condition imposed on b_j is that its probabilities are updated by Bayes' rule. That is, for any players $j, k \in \mathcal{N}$ if $\theta_j = a$ and $\theta_k = c$, then

$$
b_j(\theta_k = c \mid \theta_j = a) = \frac{b_j([\theta_j = a] \cap [\theta_k = c])}{b_j(\theta_j = a)}.
$$

(9)

Moreover, the utility function is affected by the type of each player, so now utility functions of the form $u_j(\cdot \mid \theta_j) \colon \Sigma \to \mathbb{R}$ are used, and this in turn affects the strategies, since each type of player may follow different strategies.

Now it is possible to define the expected utility for player j when facing the profile of strategies $x \in \mathcal{Q}$ as

$$
E_j(x) = \sum_{\theta_{-j} \in \Theta_{-j}} \sum_{n^1 \in \mathcal{N}} \sum_{s^1 \in \mathcal{S}_{n^1}} \cdots \sum_{n^T \in \mathcal{N}} \sum_{s^T \in \mathcal{S}_{n^T}} u_j(s^T(\theta_{n^T}), \ldots, s^1(\theta_{n^1}) \mid \theta_j)
$$

$$
\times b_j(\theta_{-j} \mid \theta_j) x_{n^T}(s^T(\theta_{n^T}) \mid s^{T-1}(\theta_{n^{T-1}}), \ldots, s^1(\theta_{n^1})) p_j(n^T)
$$

(10)

$$
\times \cdots x_{n^1}(s^1(\theta_{n^1})) p_j(n^1).
$$

3 Existence of Nash Equilibria

In this section a series of results is shown that ensures the existence of Nash equilibria in the base model of Subsect. 2.3. These results can also be adapted for each of the other models proposed to prove the existence of equilibria in them as well.

To do this, it is necessary to prove sufficient conditions of Kakutani's fixed-point theorem (see [1,5,15]) for the best response correspondence associated with the expected utility function defined for each model. The proofs are similar in all models, in some cases it is only necessary to rewrite the elements in the expected utility function, which doesn't affect the proofs themselves.

Theorem 1. *The expected utility function is a continuous function in each player's plan of conditioned strategies.*

Proof. For a given profile of conditioned strategies $x = (x_1, \ldots, x_n)$, the jth player's expected utility function can be written as follows:

$$E_j(x) = \sum_{s^1 \in S_j} \cdots \sum_{s^T \in S_j} u_j(s^T, \ldots, s^1)$$

$$\times x_j(s^T \mid s^{T-1}, \ldots, s^1) p_j(j) \cdots x_j(s^1) p_j(j)$$

$$+ \left(\sum_{n^1 \in \mathcal{N} \setminus \{j\}} \sum_{s^1 \in S_{n^1}} \sum_{s^2 \in S_j} \cdots \sum_{s^T \in S_j} u_j(s^T, \cdots, s^2, s^1) \right.$$

$$\times x_j(s^T \mid s^{T-1}, \cdots, s^2, s^1) p_j(j) \cdots x_j(s^2 \mid s^1) p_j(j) x_{n^1}(s^1) p_j(n^1)$$

$$+ \cdots + \sum_{s^1 \in S_j} \cdots \sum_{s^{T-1} \in S_j} \sum_{n^T \in \mathcal{N} \setminus \{j\}} \sum_{s^T \in S_{n^T}} u_j(s^T, s^{T-1}, \cdots, s^1) \qquad (11)$$

$$\times x_{n^T}(s^T \mid s^{T-1}, \ldots, s^1) p_j(n^T) x_j(s^{T-1} \mid s^{T-2}, \ldots, s^1)$$

$$\left. \times p_j(j) \cdots x_j(s^1) p_j(j) \right) + \cdots$$

$$+ \sum_{n^1 \in \mathcal{N} \setminus \{j\}} \sum_{s^1 \in S_{n^1}} \cdots \sum_{n^T \in \mathcal{N} \setminus \{j\}} \sum_{s^T \in S_{n^T}} u_j(s^T, \cdots, s^1)$$

$$\times x_{n^T}(s^T \mid s^{T-1}, \ldots, s^1) p_j(n^T) \cdots x_{n^1}(s^1) p_j(n^1),$$

where the first term is the sum of functions in T variables dependent of the jth player, all of which are multiplying each other, then the next bunch (noted by the parentheses) are all the sums of functions in $T - 1$ variables dependent on the jth player, and so on, with the last term being a constant function for j. Therefore the expected utility function is a continuous function in the jth player's plan of conditioned strategies. □

Theorem 2. *The set Q is a non-empty, compact and convex subset of $\mathbb{R}^q$ for a suitable q.*

Proof. For a fixed scenario $s^\ell = (s^{\ell-1}, \ldots, s^1)$, the mixed conditioned strategy of player j for scenario s^ℓ lies in a $(a_j^\ell - 1)$-simplex, where a_j^ℓ is the number of strategies available to player j at period ℓ for scenario s^ℓ. For each player j, the set of plans of mixed conditioned strategies $\mathcal{Q}_j$ is the cartesian product of the simplices for all the possible scenarios, and the set of profiles $\mathcal{Q}$ is the cartesian product of the sets of plans $\mathcal{Q}_j$. As the cartesian product of simplices, $\mathcal{Q}$ is a non-empty, compact and convex subset of $\mathbb{R}^q$ for some q.   □

Define for each player j, the best response correspondence BR_j for the partial profile x_{-j} as

$$BR_j(x_{-j}) = \{x'_j \in \mathcal{Q}_j \mid E_j(x_{-j}, x'_j) \geq E_j(x_{-j}, y_j) \text{ for all } y_j \in \mathcal{Q}_j\}. \tag{12}$$

Theorem 3. *The best response correspondence* $BR \colon \mathcal{Q} \to \mathcal{Q}$, *given by*

$$BR(x) = (BR_1(x_{-1}), BR_2(x_{-2}), \ldots, BR_N(x_{-N}), \tag{13}$$

is a non-empty correspondence with a closed graph.

Proof. Since the expected utility function is a continuous function defined on a compact set, for each player j and each $x \in Q$ it must achieve its maximum at some point $\hat{x}_j \in \mathcal{Q}_j$. Therefore, BR_j is non-empty for every player and every $x \in \mathcal{Q}$, which implies that BR is a non-empty correspondence for every $x \in \mathcal{Q}$.

Now, consider a sequence of strategy profiles $(x_{[h]})_{h=1}^\infty$, and the associated sequence of best responses $(x'_{[h]})_{h=1}^\infty$, that is, $x'_{[h]} \in BR(x_{[h]})$ for each h. Let both sequences be convergent, with $x^* = \lim_{h \to \infty} x_{[h]}$ and $x'^* = \lim_{h \to \infty} x'_{[h]}$. Fixing player j, there is $x'_{[h]_j} \in BR_j(x_{[h]_{-j}})$, which means that

$$E_j(x_{[h]_{-j}}, x'_{[h]_j}) \geq E_j(x_{[h]_{-j}}, y_j) \tag{14}$$

for any $y_j \in \mathcal{Q}_j$. As the expected utility function is continuous in each player's plan of strategies, it is possible to take limits on both sides while preserving the inequality, which means that

$$\lim_{h \to \infty} E_j(x_{[h]_{-j}}, x'_{[h]_j}) \geq \lim_{h \to \infty} (x_{[h]_{-j}}, y_j), \tag{15}$$

and interchanging the order of limits and sums,

$$E_j(x^*_{-j}, x'^*_j) \geq E_j(x^*_{-j}, y_j) \tag{16}$$

for all $y_j \in \mathcal{Q}_j$. This implies that $x'^*_j \in BR(x^*_{-j})$ for each player j, and therefore, $x'^* \in BR(x^*)$.   □

Finally, the convexity of the best response correspondence will require a few more arguments. Given a plan of mixed strategies x_j, y_j is defined as a **similar plan of strategies for the scenario** $\mathbf{s}^\ell = (\mathbf{s}^{\ell-1}, \ldots, \mathbf{s}^1)$ if y_j is a plan in which all mixed strategies are the same as in x_j, except for the one conditioned by $\mathbf{s}^\ell$, which is replaced in y_j by a pure conditioned strategy s_j^ℓ such that $x_j(s_j^\ell \mid \mathbf{s}^\ell) > 0$. That is, a strategy s_j^ℓ is chosen with positive probability in x_j when facing the scenario $\mathbf{s}^\ell$. The set of similar plans to x_j for the scenario $\mathbf{s}^\ell$ is denoted by $\mathcal{W}_j(x_j \mid \mathbf{s}^\ell)$.

Lemma 1. *Let $x \in Q$ be such that for player j, $x_j \in BR_j(x_{-j})$. Then for any scenario $\mathbf{s}^\ell = (\mathbf{s}^{\ell-1}, \ldots, \mathbf{s}^1)$ and any two plans $y_j, z_j \in \mathcal{W}_j(x_j \mid \mathbf{s}^\ell)$*

$$E_j(x_{-j}, y_j) = E_j(x_{-j}, z_j). \tag{17}$$

Proof. Fix scenario $\mathbf{s}^\ell = (\mathbf{s}^{\ell-1}, \ldots, \mathbf{s}^1)$, with $\mathbf{s}^1 \in \mathcal{S}_{n^1}, \ldots, \mathbf{s}^{\ell-1} \in \mathcal{S}_{n^{\ell-1}}$ for players $n^1, \ldots, n^{\ell-1}$ selected in the first $\ell-1$ turns. Assume that $E_j(x_{-j}, y_j) > E_j(x_{-j}, z_j)$ for some $y_j, z_j \in \mathcal{W}_j(x_j \mid \mathbf{s}^\ell)$. Observe that $E_j(x_{-j}, y_j)$ can be split as

$$
\begin{aligned}
E_j(x_{-j}, y_j) = &\sum_{s^\ell \in \mathcal{S}_j} \sum_{n^{\ell+1} \in \mathcal{N}} \sum_{s^{\ell+1} \in \mathcal{S}_{n^{\ell+1}}} \cdots \sum_{n^T \in \mathcal{N}} \sum_{s^T \in \mathcal{S}_{n^T}} u_j(s^T, \ldots, s^\ell, \mathbf{s}^\ell) \\
&\times x_{n^T}(s^T \mid s^{T-1} \cdots, s^\ell, s^{\ell-1}, \cdots, s^1) p_j(n^T) \cdots \\
&\times x_{n^{\ell+1}}(s^{\ell+1} \mid s^\ell, s^{\ell-1}, \ldots, s^1) p_j(n^{\ell+1}) y_j(s^\ell \mid s^{\ell-1}, \cdots, s^1) \\
&\times p_j(j) x_{n^{\ell-1}}(s^{\ell-1} \mid s^{\ell-2}, \ldots, s^1) p_j(n^{\ell-1}) \cdots x_{n^1}(s^1) p_j(n^1) \\
&+ \sum_{n^1 \in \mathcal{N}} \sum_{s^1 \in \mathcal{S}_{n^1}} \cdots \sum_{n^T \in \mathcal{N}} \sum_{s^T \in \mathcal{S}_{n^T}} u_j(s^T, \ldots, s^1) \\
&\times x_{n^T}(s^T \mid s^{T-1}, \ldots, s_1) p_j(n^T) \cdots x_{n^1}(s^1) p_j(n^1)
\end{aligned}
\tag{18}
$$

where in the second set of sums either $(n^1, \ldots, n^{\ell-1}, n^\ell) \neq (n^1, \ldots, n^{\ell-1}, j)$ and/or $(s^1, \ldots, s^{\ell-1}) \neq (\mathbf{s}^1, \ldots, \mathbf{s}^{\ell-1})$. An analogous expression is found for $E_j(x_{-j}, z_j)$. It is observed that y_j and z_j are only employed in the first part of (18), so it follows that

$$
\begin{aligned}
&\sum_{s^\ell \in \mathcal{S}_j} \sum_{n^{\ell+1} \in \mathcal{N}} \sum_{s^{\ell+1} \in \mathcal{S}_{n^{\ell+1}}} \cdots \sum_{n^T \in \mathcal{N}} \sum_{s^T \in \mathcal{S}_{n^T}} u_j(s^T, \ldots, s^\ell, s^{\ell-1}, \cdots s^1) \\
&\times x_{n^T}(s^T \mid s^{T-1} \cdots, s^\ell, s^{\ell-1}, \cdots, s^1) p_j(n^T) \cdots \\
&\times x_{n^{\ell+1}}(s^{\ell+1} \mid s^\ell, s^{\ell-1}, \ldots, s^1) p_j(n^{\ell+1}) y_j(s^\ell \mid s^{\ell-1}, \cdots, s^1) p_j(j) \\
&\times x_{n^{\ell-1}}(s^{\ell-1} \mid s^{\ell-2}, \ldots, s^1) p_j(n^{\ell-1}) \cdots x_{n^1}(s^1) p_j(n^1) \\
> &\sum_{s^\ell \in \mathcal{S}_j} \sum_{n^{\ell+1} \in \mathcal{N}} \sum_{s^{\ell+1} \in \mathcal{S}_{n^{\ell+1}}} \cdots \sum_{n^T \in \mathcal{N}} \sum_{s^T \in \mathcal{S}_{n^T}} u_j(s^T, \ldots, s^\ell, s^{\ell-1}, \cdots s^1) \\
&\times x_{n^T}(s^T \mid s^{T-1} \cdots, s^\ell, s^{\ell-1}, \cdots, s^1) p_j(n^T) \cdots \\
&\times x_{n^{\ell+1}}(s^{\ell+1} \mid s^\ell, s^{\ell-1}, \ldots, s^1) p_j(n^{\ell+1}) z_j(s^\ell \mid s^{\ell-1}, \cdots, s^1) p_j(j) \\
&\times x_{n^{\ell-1}}(s^{\ell-1} \mid s^{\ell-2}, \ldots, s^1) p_j(n^{\ell-1}) \cdots x_{n^1}(s^1) p_j(n^1).
\end{aligned}
\tag{19}
$$

This implies that, if s_j^ℓ and t_j^ℓ are strategies such that $y_j(s_j^\ell \mid \mathbf{s}^\ell) = 1$, and $z_j(t_j^\ell \mid \mathbf{s}^\ell) = 1$, then it is possible to replace the probability of choosing s_j^ℓ in x_j on scenario $\mathbf{s}^\ell$ with $x_j(s_j^\ell \mid \mathbf{s}^\ell) + x_j(t_j^\ell \mid \mathbf{s}^\ell)$ and the probability of choosing t_j^ℓ in x_j on scenario $\mathbf{s}^\ell$ with 0. This way, a better response to x_{-j} is obtained, compared with x_j. But this is a contradiction to the fact that x_j was a best response to x_{-j}. Therefore, $E_j(x_{-j}, y_j) = E_j(x_{-j}, z_j)$ for all $y_j, z_j \in W_j(x_j \mid \mathbf{s}^\ell)$. $\qquad \square$

Lemma 2. *Let $x \in Q$ be such that for player j, $x_j \in BR_j(x_{-j})$. Then for any scenario $\mathbf{s}^\ell = (s^{\ell-1}, \dots, \mathbf{s}^1)$ and any plan $y_j \in W_j(x_j \mid \mathbf{s}^\ell)$*

$$E_j(x) = E_j(x_{-j}, y_j). \qquad (20)$$

Proof. As in the previous lemma, $E_j(x)$ can be written in two parts

$$
E_j(x) = \sum_{s^\ell \in S_j} \sum_{n^{\ell+1} \in \mathcal{N}} \sum_{s^{\ell+1} \in S_{n^{\ell+1}}} \cdots \sum_{n^T \in \mathcal{N}} \sum_{s^T \in S_{n^T}} u_j(s^T, \dots, s^\ell, \mathbf{s}^\ell)
$$

$$
\times x_{n^T}(s^T \mid s^{T-1} \cdots, s^\ell, s^{\ell-1}, \cdots, \mathbf{s}^1) p_j(n^T) \cdots
$$

$$
\times x_{n^{\ell+1}}(s^{\ell+1} \mid s^\ell, s^{\ell-1}, \dots, \mathbf{s}^1) p_j(n^{\ell+1}) x_j(s^\ell \mid s^{\ell-1}, \cdots, \mathbf{s}^1)
$$

$$
\times p_j(j) x_{n^{\ell-1}}(s^{\ell-1} \mid s^{\ell-2}, \dots, \mathbf{s}^1) p_j(n^{\ell-1}) \cdots x_{n^1}(\mathbf{s}^1) p_j(n^1) \qquad (21)
$$

$$
+ \sum_{n^1 \in \mathcal{N}} \sum_{s^1 \in \mathcal{S}_{n^1}} \cdots \sum_{n^T \in \mathcal{N}} \sum_{s^T \in \mathcal{S}_{n^T}} u_j(s^T, \dots, s^1)
$$

$$
\times x_{n^T}(s^T \mid s^{T-1}, \dots, s_1) p_j(n^T) \cdots x_{n^1}(\mathbf{s}^1) p_j(n^1)
$$

where in the second set of sums either $(n^1, \dots, n^{\ell-1}, n^\ell) \neq (n^1, \dots, n^{\ell-1}, j)$ and/or $(s^1, \dots, s^{\ell-1}) \neq (\mathbf{s}^1, \dots, \mathbf{s}^{\ell-1})$ holds. The first part of the previous expression can be written as

$$
\sum_{z_j \in W_j(x_j \mid \mathbf{s}^\ell)} \sum_{s^\ell \in S_j} \sum_{n^{\ell+1} \in \mathcal{N}} \sum_{s^{\ell+1} \in S_{n^{\ell+1}}} \cdots \sum_{n^T \in \mathcal{N}} \sum_{s^T \in S_{n^T}} u_j(s^T, \dots, s^\ell, \mathbf{s}^\ell)
$$

$$
\times x_{n^T}(s^T \mid s^{T-1} \cdots, s^\ell, s^{\ell-1}, \cdots, \mathbf{s}^1) p_j(n^T) \cdots \qquad (22)
$$

$$
\times x_{n^{\ell+1}}(s^{\ell+1} \mid s^\ell, s^{\ell-1}, \dots, \mathbf{s}^1) p_j(n^{\ell+1}) x_j(s^\ell \mid s^{\ell-1}, \cdots, \mathbf{s}^1) z_j(s^\ell \mid \mathbf{s})
$$

$$
\times p_j(j) x_{n^{\ell-1}}(s^{\ell-1} \mid s^{\ell-2}, \dots, \mathbf{s}^1) p_j(n^{\ell-1}) \cdots x_{n^1}(\mathbf{s}^1) p_j(n^1)
$$

that is, as the weighted sum of the first part of the expressions of the form (18) for every $z_j \in W_j(x_j \mid \mathbf{s}^\ell)$, with weights $x_j(s_j \mid \mathbf{s}^\ell)$ where s_j is chosen as the strategy in scenario $\mathbf{s}^\ell$ so that $z_j(s_j \mid \mathbf{s}^\ell) = 1$ for each z_j. Weighing the whole expression (18) with each z_j accordingly, the expected utility of profile x can be written as the weighted sum of the expected utilities of all similar plans in

$W_j(x_j \mid \mathbf{s}^\ell)$. In Lemma 1, it is shown that all expected utilities are equal for all similar plans, which means that each expected utility can be replaced with the expected utility for a particular similar plan y_j. Since the weights are chosen so that their sum is 1, the equality stated above follows.   □

From the previous result an easy corollary follows, which is obtained by applying the previous result to each pure strategy in the mixed plan y_j.

Corollary 1. *Let $x \in Q$ be such that for player j, $x_j \in BR_j(x_{-j})$. For any scenario $\mathbf{s}^\ell = (\mathbf{s}^{\ell-1}, \ldots, \mathbf{s}^1)$, if the plan of mixed conditioned strategies y_j on scenario $\mathbf{s}^\ell$ is such that $\mathcal{W}_j(y_j \mid \mathbf{s}^\ell) \subseteq \mathcal{W}_j(x_j \mid \mathbf{s}^\ell)$, then*

$$E_j(x) = E_j(x_{-j}, y_j). \tag{23}$$

This, in turn, provides the next result by applying Corollary 1 twice to the mixed plans y_j and z_j.

Corollary 2. *Let $x \in Q$ be such that for player j, $x_j \in BR_j(x_{-j})$. For any scenario $\mathbf{s}^\ell = (\mathbf{s}^{\ell-1}, \ldots, \mathbf{s}^1)$, if the plans of mixed conditioned strategies y_j and z_j on scenario $\mathbf{s}^\ell$ are such that $\mathcal{W}_j(y_j \mid \mathbf{s}^\ell) \subseteq \mathcal{W}_j(x_j \mid \mathbf{s}^\ell)$ and $\mathcal{W}_j(z_j \mid \mathbf{s}^\ell) \subseteq \mathcal{W}_j(x_j \mid \mathbf{s}^\ell)$, then*

$$E_j(x_{-j}, y_j) = E_j(x_{-j}, z_j). \tag{24}$$

It is possible to generalize the previous results in the following theorem.

Theorem 4. *Let $x \in Q$ be such that for player j, $x_j \in BR_j(x_{-j})$. Then, for any scenario $\mathbf{s}^\ell = (\mathbf{s}^{\ell-1}, \ldots, \mathbf{s}^1)$ and any two plans of strategies y_j, z_j such that $\mathcal{W}_j(y_j \mid \mathbf{s}^\ell) \subseteq \mathcal{W}_j(x_j \mid \mathbf{s}^\ell)$ and $\mathcal{W}_j(z_j \mid \mathbf{s}^\ell \subseteq \mathcal{W}_j(x_j \mid \mathbf{s}^\ell)$*

$$E_j(x_{-j}, y_j) = E_j(x_{-j}, z_j). \tag{25}$$

Also, by noting that if x_j, x_j' are best responses to x_{-j}, then $E_j(x_{-j}, x_j) = E_j(x_{-j}, x_j')$, and therefore the plan $x_j'' = (x_j + x_j')/2$ is a best response to x_{-j}, and the next result follows.

Theorem 5. *Let x_{-j} be a partial profile of plans of strategies. If y_j, z_j are plans of strategies such that for scenario $\mathbf{s}^\ell = (\mathbf{s}^{\ell-1}, \ldots, \mathbf{s}^1)$ there exist $x_j, x_j' \in BR_j(x_{-j})$ that meet the condition $\mathcal{W}_j(y_j \mid \mathbf{s}^\ell) \subseteq \mathcal{W}_j(x_j \mid \mathbf{s}^\ell)$ and $\mathcal{W}_j(z_j \mid \mathbf{s}^\ell) \subseteq \mathcal{W}_j(x_j' \mid \mathbf{s}^\ell)$, then it holds that*

$$E_j(x_{-j}, y_j) = E_j(x_{-j}, z_j). \tag{26}$$

With the previous results, the last condition needed for the best response correspondence can be easily proved.

Theorem 6. *The best response correspondence BR is convex.*

Proof. As it has been observed above, given x_j, x_j' are best responses to a partial profile x_{-j}, it holds that any convex combination of x_j and x_j' is a best response, since $E_j(x_{-j}, x_j) = E_j(x_{-j}, x_j')$, then for $x_j'' = \lambda x_j + (1-\lambda)x_j'$, with $\lambda \in [0,1]$, it follows that $E_j(x_{-j}, x_j'') = E_j(x_{-j}, x_j)$.   □

The previous discussion shows that the best response correspondence meets the conditions of Kakutani's fixed-point theorem, which guarantees the existence of at least one fixed point for BR. It can easily be seen that a fixed point for BR corresponds to a Nash equilibria in the model, and vice versa. This way, the central theorem of the work is obtained.

Theorem 7. *Every sequential game with finite horizon and turn selection process with finite strategies sets has at least one Nash equilibrium.*

4 Examples

4.1 Adding Ones and Zeros

Two players decide to engage in the following game: each turn a coin is flipped to select one of the two players, say, if the coin shows heads, then player 1 is selected, and if the coin shows tails, then player 2 is selected. Once this has happened, the selected player can choose whether he should add one or zero to the counter of the game. These steps are repeated for T turns. The counter starts at zero, and shows the amount of money that one player receives from the other, depending on the parity: if the counter is odd, then 1 receives the amount from 2; if the counter is even, then 2 receives the amount from 1.

The game proposed above will be studied for the cases where $T = 2, 3$. First, it is noticed that in both cases, the player who decides the winner is the one that is selected in the last turn, since he can change the parity of the counter with his decision. The decisions of the players selected before only influence the amount of money in play.

In the case where $T = 2$, if the strategies sets for each player are $\mathcal{S}_1 = \mathcal{S}_2 = \{0, 1\}$, then

$$
\begin{aligned}
x_1(0 \mid 0) &= 0 \quad x_1(1 \mid 0) = 1 \\
x_1(0 \mid 1) &= 1 \quad x_1(1 \mid 1) = 0 \\
x_2(0 \mid 0) &= 1 \quad x_2(1 \mid 0) = 0 \\
x_2(0 \mid 1) &= 0 \quad x_2(1 \mid 1) = 1
\end{aligned}
\tag{27}
$$

which can be summarized as

$$
x_1(s^2 \mid s^1) = \frac{1}{2} + \frac{(-1)^{s^1 + s^2 + 1}}{2}
\tag{28}
$$

$$
x_2(s^2 \mid s^1) = \frac{1}{2} + \frac{(-1)^{s^1 + s^2}}{2}.
$$

The utility functions for the players are

$$
u_1(s^2, s^1) = (-1)^{s^1 + s^2 + 1}(s^1 + s^2)
\tag{29}
$$

and

$$
u_2(s^2, s^1) = (-1)^{s^1 + s^2}(s^1 + s^2).
\tag{30}
$$

After some manipulations, the expected utility for player 1 is

$$E_1(x) = p^2 - 2pqx_1(1) + pq - 2q^2x_2(1),$$ (31)

where p and $q = 1 - p$ are the probabilities of the coin showing heads and tails, respectively. For player 2, it can be easily seen that $E_2(x) = E_1(x)$, since the winnings of one player are the losses of the other player. To maximise their respective expected utility, player 1 follows the strategy $x_1(1) = 0$, and player 2 follows the strategy $x_2(1) = 1$. That is, if player 1 is selected in the first turn, he chooses to keep the counter in zero, whereas if player 2 is selected, he chooses to increase the counter to one.

Now, if the horizon of the game is $T = 3$ the following strategies for the last turn of the game are obtained, given by

$$x_1(s^3 \mid s^2, s^1) = \frac{1}{2} + \frac{(-1)^{s^1+s^2+s^3+1}}{2}$$ (32)

and

$$x_2(s^3 \mid s^2, s^1) = \frac{1}{2} + \frac{(-1)^{s^1+s^2+s^3}}{2}.$$ (33)

There are also the corresponding utility functions

$$u_1(s^3, s^2, s^1) = (-1)^{s^1+s^2+s^3+1}(s^1 + s^2 + s^3)$$ (34)

and

$$u_2(s^3, s^2, s^1) = (-1)^{s^1+s^2+s^3}(s^1 + s^2 + s^3) = -u_1(s^3, s^2, s^1).$$ (35)

The expected utility function for player 1 is given by

$$\begin{aligned}
E_1(x) = {} & (1 - x_1(1))p^2 + (1 - x_2(1))pq \\
& - 2(x_1(1 \mid 0)pq + x_2(1 \mid 0)q^2)((1 - x_1(1))p + (1 - x_2(1))q) \\
& + ((1 + 2x_1(1 \mid 1))p^2 + (1 + 2x_2(1 \mid 1))pq)(x_1(1)p + x_2(1)q) \\
& - 2x_1(1)pq - 2x_2(1)q^2,
\end{aligned}$$ (36)

where the terms concerning $x_1(1 \mid 0)$ are non-positive, so to maximise (36) it is possible to take $x_1(1 \mid 0) = 0$. Since $E_2(x) = -E_1(x)$, the terms concerning $x_2(1 \mid 0)$ are non-negative in $E_2(x)$, so $x_2(1 \mid 0) = 1$ can be taken to maximise the expected utility. Now (36) becomes

$$\begin{aligned}
E_1(x) = {} & (1 - x_1(1))p^2 + (1 - x_2(1))pq \\
& - 2q^2((1 - x_1(1))p + (1 - x_2(1))q) \\
& + ((1 + 2x_1(1 \mid 1))p^2 + (1 + 2x_2(1 \mid 1))pq)(x_1(1)p + x_2(1)q) \\
& - 2x_1(1)pq - 2x_2(1)q^2,
\end{aligned}$$ (37)

and reasoning as above, it is possible to take $x_1(1 \mid 1) = 1$, $x_2(1 \mid 1) = 0$, making the expected utility of player 1

$$\begin{aligned}
E_1(x) = {} & (1 - x_1(1))p^2 + (1 - x_2(1))pq \\
& - 2q^2((1 - x_1(1))p + (1 - x_2(1))q) \\
& + (3p^2 + pq)(x_1(1)p + x_2(1)q) - 2x_1(1)pq - 2x_2(1)q^2,
\end{aligned}$$ (38)

which after many calculations, can be written as

$$E_1(x) = p - 2q^2 + (2px_1(1) + 2qx_2(1))(p - q). \tag{39}$$

Therefore $x_1(1)$ and $x_2(1)$ can be chosen according to the relation between p and q:

1. If $p < q$, then $x_1(1) = 0$ and $x_2(1) = 1$.
2. If $p > q$, then $x_1(1) = 1$ and $x_2(1) = 0$.
3. If $p = q$, then $x_1(1)$ and $x_2(1)$ can be any values in $[0, 1]$.

In this example, it is possible to see the importance of the order in which players are being selected, and the effect of the distribution chosen by each player to represent the turn selection process on the strategies that he should follow.

4.2   An Application of the Models: Picking Teammates

In the following example an application of some of the models analysed in the previous sections is shown. To be precise, the example is a combination of the models in Subsects. 2.4 and 2.6, since the probabilities of the turn selection process and the strategy sets for each period change according to the decisions made before.

In a certain sports league, every year the two teams that are involved have to choose between certain college athletes to become part of their team. This year they have to pick between athlete A and athlete B. Both teams know that choosing A is equivalent to 1 unit of utility, whereas choosing B is equivalent to 2 units of utility. The final utility obtained after the picks have been made is equal to the sum of the utilities given by the athletes chosen for the team.

However, instead of the usual picking process where the lowest ranked team picks first and then the others follow in inverse order of ranking, the sports league has devised the following system: in each of the two picking periods, each team has a positive probability of being selected to make the next choice. Since team 1 was the best ranked team, it has a probability of being selected in the first period of $1/3$, whereas team 2, being the lowest ranked has a probability of being selected in the first period of $2/3$. For the selection in the second period, the probabilities change according to the previous pick. If the team selected in the first period chooses athlete A, then for the second period its probability of being selected is reduced to half of the probability it had in the first period; if the team selected in the first period chooses B, then for the second period its probability of being selected is reduced to a third of the probability it had in the first period. In both cases, the reduced probability is added to the other team. With these new probabilities, the team is selected for the second round, and in this case, that team automatically chooses the athlete that is left.

To find the optimal decision for each team, first the probabilities of selection are computed. To do so, the strategy sets of each player are denoted by $S_1 = \{A_1, B_1\}$ and $S_2 = \{A_2, B_2\}$ where the subscripts are used only to identify the team that is choosing, so essentially A_1 and A_2 mean to pick athlete A, and B_1

and B_2 mean to pick athlete B. Since both players know the probabilities of the turn selection process, it is possible write $p_1(\cdot) = p_2(\cdot) = p(\cdot)$, with p given by

$$
\begin{aligned}
p(1) &= 1/3 & p(2) &= 2/3 \\
p(1 \mid A_1) &= 1/6 & p(2 \mid A_1) &= 5/6 \\
p(1 \mid B_1) &= 1/9 & p(2 \mid B_1) &= 8/9 \\
p(1 \mid A_2) &= 2/3 & p(2 \mid A_2) &= 1/3 \\
p(1 \mid B_2) &= 7/9 & p(2 \mid B_2) &= 2/9
\end{aligned}
\tag{40}
$$

The expected utility function is calculated for each team, inputting the probabilities for each possible scenario, and the utilities received in each case. Observe that since the second pick is automatic, $x_j(A_j \mid B_k) = x_j(B_j \mid A_k) = 1$ for all $j, k \in \{1, 2\}$. This way, the expected utilities are

$$
E_1(x) = \frac{25}{27} + \frac{1}{27}x_1(A_1) + \frac{10}{27}x_2(A_2)
\tag{41}
$$

and

$$
E_2(x) = \frac{16}{9} + \frac{7}{27}x_1(A_1) - \frac{10}{27}x_2(A_2)
\tag{42}
$$

Therefore, to maximise their expected utility functions, team 1 should follow the strategy $x_1(A_1) = 1$ and team 2 should follow the strategy $x_2(A_2) = 0$. That is, team 1 should pick A if they are selected on the first round, whereas team 2 should pick B if they are selected on the first round.

5   Conclusions

The work presented introduces a different approach to model certain situations, where the individuals involved may not know the moment in which they make a decision, or the order in which these decisions are made. These models can also work to allow the creation of situations in which players that are not equal can be equalized to a certain level, to allow a fairer competition. The ideas here presented comprise a large variety of situations, as many components of the model are allowed to change in various ways, to adapt better to the situation being studied.

The examples presented here show that players choose in a different way than expected in other models, and that these decisions may even be dependent on the way the turn selection process is modelled by each player. The examples are not made with larger sets of strategies or a larger horizon since the calculations behind them become too messy very quickly. Therefore, some future work will be focused on studying these models in their description as optimization problems, to be able to find solutions at least in an approximate form. This, in turn, implies the study of sequences of approximate solutions to study the convergence to actual equilibria of the game.

Acknowledgements. This work was partially supported by CONACYT (Mexico) and ASCR (Czech Republic) under Grants No. 171396 and 283640.

References

1. Aliprantis, C.D., Border, K.C.: Infinite Dimensional Analysis: A Hitchhiker's Guide. Springer, Heidelberg (2007)
2. Balbus, Ł., Nowak, A.S.: Existence of perfect equilibria in a class of multigenerational stochastic games of capital accumulation. Automatica **44**(6), 1471–1479 (2008)
3. Balbus, Ł., Reffett, K., Woźny, Ł.: A constructive geometrical approach to the uniqueness of Markov stationary equilibrium in stochastic games of intergenerational altruism. J. Econ. Dyn. Control **37**(5), 1019–1039 (2013)
4. Becerril-Borja, R., Montes-de-Oca, R.: Sequential games with finite horizon and turn selection process: finite strategy sets case. In: Proceedings of the 5th International Conference on Operations Research and Enterprise Systems (ICORES 2016), pp. 44–50 (2016)
5. Border, K.C.: Fixed Point Theorems with Applications to Economics and Game Theory. Cambridge University Press, Cambridge (1985)
6. Fudenberg, D., Tirole, J.: Game Theory. The MIT Press, Cambridge (1991)
7. González-Díaz, J., García-Jurado, I., Fiestras-Janeiro, M.G.: An Introductory Course on Mathematical Game Theory. Graduate Studies in Mathematics. American Mathematical Society, Providence (2010)
8. Kitti, M.: Conditionally stationary equilibria in discounted dynamic games. Dyn. Games Appl. **1**(4), 514–533 (2011)
9. Neyman, A., Sorin, S. (eds.): Stochastic Games and Applications. Springer, New York (2003)
10. Nowak, A.S.: On perfect equilibria in stochastic models of growth with intergenerational altruism. Econ. Theor. **28**(1), 73–83 (2006)
11. Nowak, A.S.: On a noncooperative stochastic game played by internally cooperating generations. J. Optim. Theor. Appl. **144**(1), 88–106 (2010)
12. Saez-Marti, M., Weibull, J.W.: Discounting and altruism to future decision makers. J. Econ. Theor. **122**(2), 254–266 (2005)
13. Tadelis, S.: Game Theory. Princeton University Press, Princeton (2013)
14. Woźny, Ł., Growiec, J.: Intergenerational interactions in human capital accumulation. BE J. Theor. Econ. **12**(1), 1–47 (2012)
15. Zeidler, E.: Nonlinear Functional Analysis and its Applications I: Fixed-Point Theorems. Springer, New York (1986)

Solving Chance-Constrained Games Using Complementarity Problems

Vikas Vikram Singh[1(✉)], Oualid Jouini[2], and Abdel Lisser[1]

[1] Laboratoire de Recherche en Informatique, Université Paris Sud XI,
Bât 650, 91405 Orsay, France
{vikas.singh,abdel.lisser}@lri.fr
[2] Laboratoire Génie Industriel, Ecole Centrale Paris,
Grande Voie des Vignes, 92290 Châtenay-Malabry, France
oualid.jouini@ecp.fr

Abstract. In this paper, we formulate the random bimatrix game as a chance-constrained game using chance constraint. We show that a Nash equilibrium problem, corresponding to independent normally distributed payoffs, is equivalent to a nonlinear complementarity problem. Further if the payoffs are also identically distributed, a strategy pair where each player's strategy is the uniform distribution over his action set, is a Nash equilibrium. We show that a Nash equilibrium problem corresponding to independent Cauchy distributed payoffs, is equivalent to a linear complementarity problem.

Keywords: Chance-constrained game · Nash equilibrium · Normal distribution · Cauchy distribution · Nonlinear complementarity problem · Linear complementarity problem

1 Introduction

It is well known that there exists a mixed strategy saddle point equilibrium for a two player zero sum matrix game [22]. John Nash [21] showed the existence of a mixed strategy equilibrium for the games with finite number of players where each player has finite number of actions. Later such equilibrium was called Nash equilibrium. For two player case the game considered in [21] can be represented by $m \times n$ matrices A and B. The matrices $A = [a_{ij}]$ and $B = [b_{ij}]$ denote the payoff matrices of player 1 and player 2 respectively, and m, n denote the number of actions of player 1 and player 2 respectively. Let $I = \{1, 2, \cdots, m\}$, and $J = \{1, 2, \cdots, n\}$ be the action sets of player 1 and player 2 respectively. The sets I and J are also called the sets of pure strategies of player 1 and player 2 respectively. The set of mixed strategies of each player is defined by the set of all probability distributions over his action set. Let $X = \{x = (x_1, x_2, \cdots, x_m) | \sum_{i=1}^{m} x_i = 1, x_i \geq 0, \forall\ i \in I\}$ and $Y = \{y = (y_1, y_2, \cdots, y_n) | \sum_{j=1}^{n} y_j = 1, y_j \geq 0, \forall\ j \in J\}$ be the sets of mixed strategies of player 1 and player 2 respectively. For a given strategy pair (x, y),

© Springer International Publishing AG 2017
B. Vitoriano and G.H. Parlier (Eds.): ICORES 2016, CCIS 695, pp. 52–67, 2017.
DOI: 10.1007/978-3-319-53982-9_4

the payoffs of player 1 and player 2 are given by $x^T A y$ and $x^T B y$ respectively; T denotes the transposition. For a fixed strategy of one player, another player seeks for a strategy that gives him the highest payoff among all his other strategies. Such a strategy is called the best response strategy. The set of best response strategies of player 1 for a fixed strategy y of player 2 is given by

$$BR(y) = \left\{ \bar{x} | \bar{x}^T A y \geq x^T A y, \; \forall \, x \in X \right\}.$$

The set of best response strategies of player 2 for a fixed strategy x of player 1 is given by

$$BR(x) = \left\{ \bar{y} | x^T B \bar{y} \geq x^T B y, \; \forall \, y \in Y \right\}.$$

A strategy pair (x^*, y^*) is said to be a Nash equilibrium if and only if $x^* \in BR(y^*)$ and $y^* \in BR(x^*)$. A Nash equilibrium of above bimatrix game can be obtained by solving a linear complementarity problem (LCP) [16].

Both [21] and [22] considered the games where the payoffs of players are exact real values. In some cases the payoffs of players may be within certain ranges. In [7] these situations are modeled as interval valued matrix game using fuzzy theory. The computational approaches have been proposed to solve interval valued matrix game (see [10,17,19]). However, in many situations payoffs are random variables due to uncertainty which arises from various external factors. The wholesale electricity markets are the good examples (see [8,18,30,31]). One way to handle this type of game is by taking the expectation of random payoffs and consider the corresponding deterministic game (see [30,31]). Some recent papers on the games with random payoffs using expected payoff criterion include [9,12,25,32].

The expected payoff criterion does not take a proper account of stochasticity in the cases where the observed sample payoffs are large amounts with very small probabilities. These situations are better handled by considering a payoff criterion based on chance constraint programming (see [4,6,24]). In this payoff criterion, the payoff of a player is defined using a chance constraint and for this reason these games are called chance-constrained games. There are few papers on zero sum chance-constrained games available in the literature (see [2,3,5,29]). Recently, a chance-constrained game with finite number of players is considered in [27,28] where authors showed the existence of a mixed strategy Nash equilibrium. In [27], the case where the random payoff vector of each player follow a certain distribution is considered. In particular, the authors considered the case where the components of the payoff vector of each player are independent normal/Cauchy random variables, and they also consider the case where the payoff vector of each player follow a multivariate elliptically symmetric distribution. In [28], the case where the distribution of payoff vector of each player is not known completely is considered. The authors consider a distributionally robust approach to handle these games. In application regimes some chance-constrained game models have been considered, e.g., see [8,18]. In [18], the randomness in payoffs is due to the installation of wind generators on electricity market, and they consider the case of independent normal random variables. Later, for better representation and ease in computation the authors, in detail, considered the case where only one wind generator is installed in the electricity market. In [8], the payoffs are random due to uncertain demand from consumers which is assumed to be normally distributed.

In this paper, we consider the case where the entries of the payoff matrices are independent random variables following the same distribution (possibly with different parameters). For a given strategy pair (x, y), the payoff of each player is a random variable which is a linear combination of the independent random variables. We consider the distributions that are closed under a linear combination of the independent random variables. The normal and Cauchy distributions satisfy this property. We consider each distribution separately. We show that a Nash equilibrium of the chance-constrained game corresponding to normal distribution can be obtained by solving an equivalent nonlinear complementarity problem (NCP). Further we consider a special case where the entries of the payoff matrices are also identically distributed. We show that a strategy pair, where each player's strategy is a uniform distribution over his action set, is a Nash equilibrium. We show that a Nash equilibrium of the chance-constrained game corresponding to Cauchy distribution can be obtained by solving an equivalent LCP.

Now, we describe the structure of rest of the paper. Section 2 contains the definition of a chance-constrained game. Section 3 contains the complementarity problem formulation of chance-constrained game. We conclude the paper in Sect. 4

2   The Model

We consider two player bimatrix game where the entries of the payoff matrices are random variables. We denote the random payoff matrices of player 1 and player 2 by A^w and B^w respectively, where w denotes the uncertainty parameter. Let $(\Omega, \mathcal{F}, P)$ be a probability space. Then, for each $i \in I$, $j \in J$, $a_{ij}^w : \Omega \to \mathbb{R}$, and $b_{ij}^w : \Omega \to \mathbb{R}$. For each $(x, y) \in X \times Y$, the payoffs $x^T A^w y$ and $x^T B^w y$ of player 1 and player 2 respectively would be random variables. For a strategy pair (x, y), each player is interested in the highest level of his payoff that can be attained with at least a specified level of confidence. The confidence level of each player is given a priori. We assume that the confidence level of one player is known to another player. Let $\alpha_1 \in [0, 1]$ and $\alpha_2 \in [0, 1]$ be the confidence levels of player 1 and player 2 respectively. Let $\alpha = (\alpha_1, \alpha_2)$ be a confidence level vector. For a given strategy pair (x, y) and a given confidence level vector α, the payoff of player 1 is given by

$$u_1^{\alpha_1}(x, y) = \sup\{u | P(x^T A^w y \geq u) \geq \alpha_1\}, \tag{1}$$

and the payoff of player 2 is given by

$$u_2^{\alpha_2}(x, y) = \sup\{v | P(x^T B^w y \geq v) \geq \alpha_2\}. \tag{2}$$

We assume that the probability distributions of the entries of the payoff matrix of one player are known to another player. Then, for a given α the payoff function of one player defined above is known to another player. That is, for a given α the chance-constrained game is a non-cooperative game with complete information. For a given α, the set of best response strategies of player 1 against the fixed strategy y of player 2 is given by

$$BR^{\alpha_1}(y) = \{\bar{x} \in X | u_1^{\alpha_1}(\bar{x}, y) \geq u_1^{\alpha_1}(x, y), \ \forall \ x \in X\},$$

and the set of best response strategies of player 2 against the fixed strategy x of player 1 is given by

$$BR^{\alpha_2}(x) = \{\bar{y} \in Y | u_2^{\alpha_2}(x, \bar{y}) \geq u_2^{\alpha_2}(x, y), \ \forall \ y \in Y\}.$$

Definition 1 (Nash Equilibrium). *For a given confidence level vector α, a strategy pair (x^*, y^*) is said to be a Nash equilibrium of the chance-constrained game if the following inequalities hold:*

$$u_1^{\alpha_1}(x^*, y^*) \geq u_1^{\alpha_1}(x, y^*), \ \forall \ x \in X,$$
$$u_2^{\alpha_2}(x^*, y^*) \geq u_2^{\alpha_2}(x^*, y). \ \forall \ y \in Y.$$

3 Complementarity Problem for Chance-Constrained Game

In this section, we consider the case where the entries of payoff matrix A^w of player 1 are independent random variables following a certain distribution, and the entries of payoff matrix B^w of player 2 are independent random variables following a certain distribution. Then, at strategy pair (x, y) the payoff of each player is a linear combination of the independent random variables. We are interested in those probability distributions that are closed under a linear combination of the independent random variables. That is, if $Y_1, Y_2, \cdots, Y_k$ are independent random variables following the same distribution (possibly with different parameters), for any $b \in \mathbb{R}^k$, the distribution of $\sum_{i=1}^{k} b_i Y_i$ is same as Y_i up to parameters. The normal and Cauchy distributions satisfy the above property [13]. For the case of normal distribution we show that a Nash equilibrium of the chance-constrained game can be obtained by solving an equivalent NCP, and for the case of Cauchy distribution we show that a Nash equilibrium of the chance-constrained game can be obtained by solving an equivalent LCP.

3.1 Payoffs Following Normal Distribution

We assume that all the components of matrix A^w are independent normal random variables, where the mean and variance of a_{ij}^w, $i \in I$, $j \in J$, are $\mu_{1,ij}$ and $\sigma_{1,ij}^2$ respectively, and all the components of matrix B^w are independent normal random variables, where the mean and variance of b_{ij}^w, $i \in I$, $j \in J$, are $\mu_{2,ij}$ and $\sigma_{2,ij}^2$ respectively. For a given strategy pair (x, y), $x^T A^w y$ follows a normal distribution with mean $\mu_1(x, y) = \sum_{i \in I, j \in J} \mu_{1,ij} x_i y_j$ and variance $\sigma_1^2(x, y) = \sum_{i \in I, j \in J} x_i^2 y_j^2 \sigma_{1,ij}^2$, and $x^T B^w y$ follows a normal distribution with mean $\mu_2(x, y) = \sum_{i \in I, j \in J} \mu_{2,ij} x_i y_j$ and variance $\sigma_2^2(x, y) = \sum_{i \in I, j \in J} x_i^2 y_j^2 \sigma_{2,ij}^2$. Then, $Z_1^N = \frac{x^T A^w y - \mu_1(x, y)}{\sigma_1(x, y)}$ and $Z_2^N = \frac{x^T B^w y - \mu_2(x, y)}{\sigma_2(x, y)}$ follow a standard normal distribution. Let $F_{Z_1^N}^{-1}(\cdot)$ and $F_{Z_2^N}^{-1}(\cdot)$ be the quantile functions of a standard normal distribution. From (1), for a given strategy pair (x, y) and a given confidence level α_1, the payoff of player 1 is given by

$$
\begin{aligned}
u_1^{\alpha_1}(x, y) &= \sup\{u | P(x^T A^w y \ge u) \ge \alpha_1\} \\
&= \sup\{u | P(x^T A^w y \le u) \le 1 - \alpha_1\} \\
&= \sup\left\{u \Big| F_{Z_1^N}\left(\frac{u - \mu_1(x, y)}{\sigma_1(x, y)}\right) \le 1 - \alpha_1\right\} \\
&= \sup\left\{u | u \le \mu_1(x, y) + \sigma_1(x, y) F_{Z_1^N}^{-1}(1 - \alpha_1)\right\}.
\end{aligned}
$$

That is,

$$
u_1^{\alpha_1}(x, y) = \sum_{i \in I, j \in J} \mu_{1,ij} x_i y_j + \left(\sum_{i \in I, j \in J} x_i^2 y_j^2 \sigma_{1,ij}^2\right)^{1/2} F_{Z_1^N}^{-1}(1 - \alpha_1). \tag{3}
$$

Similarly, from (2) for a given strategy pair (x, y) and a given confidence level α_2, the payoff of player 2 is given by

$$
u_2^{\alpha_2}(x, y) = \sum_{i \in I, j \in J} \mu_{2,ij} x_i y_j + \left(\sum_{i \in I, j \in J} x_i^2 y_j^2 \sigma_{2,ij}^2\right)^{1/2} F_{Z_2^N}^{-1}(1 - \alpha_2). \tag{4}
$$

Theorem 1. *Consider a bimatrix game (A^w, B^w). If all the components of matrix A^w are independent normal random variables, and all the components of matrix B^w are also independent normal random variables, there exists a mixed strategy Nash equilibrium for chance-constrained game for all $\alpha \in [0.5, 1]^2$.*

Proof. The proof follows from [27].

Nonlinear Complementarity Problem Formulation. The payoff function of player 1 defined by (3) can be written as follows:

$$
u_1^{\alpha_1}(x, y) = x^T \mu_1(y) + ||\Sigma_1^{1/2}(y) x|| F_{Z_1^N}^{-1}(1 - \alpha_1), \tag{5}
$$

where $|| \cdot ||$ is the Euclidean norm, and $\mu_1(y) = (\mu_{1,i}(y))_{i \in I}$ is an $m \times 1$ vector where $\mu_{1,i}(y) = \sum_{j \in J} \mu_{1,ij} y_j$, and $\Sigma_1(y)$ is an $m \times m$ diagonal matrix whose ith diagonal entry $\Sigma_{1,ii}(y) = \sum_{j \in J} \sigma_{1,ij}^2 y_j^2$. Similarly, the payoff function of player 2 defined by (4) can be written as follows:

$$
u_2^{\alpha_2}(x, y) = y^T \mu_2(x) + ||\Sigma_2^{1/2}(x) y|| F_{Z_2^N}^{-1}(1 - \alpha_2), \tag{6}
$$

where $\mu_2(x) = \left(\mu_{2,j}(x)\right)_{j \in J}$ is an $n \times 1$ vector where $\mu_{2,j}(x) = \sum_{i \in I} \mu_{2,ij} x_i$, and $\Sigma_2(x)$ is an $n \times n$ diagonal matrix whose jth diagonal entry $\Sigma_{2,jj}(x) = \sum_{i \in I} \sigma_{2,ij}^2 x_i^2$. For fixed $y \in Y$ and $\alpha_1 \in [0.5, 1]$, the payoff function $u_1^{\alpha_1}(\cdot, y)$ of player 1 defined by (5) is a concave function of x because $F_{Z_1^N}^{-1}(1 - \alpha_1) \le 0$ for all $\alpha_1 \in [0.5, 1]$. Similarly, for fixed $x \in X$ and $\alpha_2 \in [0.5, 1]$, the payoff function $u_2^{\alpha_2}(x, \cdot)$ of player 2 defined by (6) is a concave function of y.

Then, for a fixed $y \in Y$ and $\alpha_1 \in [0.5, 1]$, a best response strategy of player 1 can be obtained by solving the convex quadratic program [QP1] given below:

$$[\text{QP1}] \quad \max_x \quad x^T \mu_1(y) + ||\Sigma_1^{1/2}(y)x|| F_{Z_1^N}^{-1}(1 - \alpha_1)$$

$$\text{s.t.}$$

$$(i) \sum_{i \in I} x_i = 1,$$

$$(ii) \; x_i \ge 0, \; i \in I.$$

It is easy to see that a feasible solution of [QP1] satisfies the linear independence constraint qualification. Then, Karush-Kuhn-Tucker (KKT) conditions of [QP1] will be necessary and sufficient conditions for optimal solution (for details see [1,23]). For a given vector ν, $\nu \ge 0$ means $\nu_k \ge 0$, for all k. The equality constraint of [QP1] can be replaced by two equivalent inequality constraints, and the free Lagrange multiplier corresponding to equality constraint can be replaced by the difference of two nonnegative variables. By using these transformations, the best response strategy of player 1 can be obtained by solving the following KKT conditions of [QP1]:

$$(7) \quad \begin{cases} 0 \le x \perp -\mu_1(y) - \dfrac{\Sigma_1(y)x \cdot c_{\alpha_1}}{||\Sigma_1^{1/2}(y)x||} - \lambda_1 e_m + \lambda_2 e_m \ge 0, \\ 0 \le \lambda_1 \perp \sum_{i \in I} x_i - 1 \ge 0, \\ 0 \le \lambda_2 \perp 1 - \sum_{i \in I} x_i \ge 0, \end{cases}$$

where e_m is the $m \times 1$ vector of ones, and $c_{\alpha_1} = F_{Z_1^N}^{-1}(1 - \alpha_1)$, and $\perp$ means that elementwise equality must hold at one or both sides. For fixed $x \in X$ and $\alpha_2 \in [0.5, 1]$, a best response strategy of player 2 can be obtained by solving the convex quadratic program [QP2] given below:

$$[\text{QP2}] \quad \max_y \quad y^T \mu_2(x) + ||\Sigma_2^{1/2}(x)y|| F_{Z_2^N}^{-1}(1 - \alpha_2)$$

$$\text{s.t.}$$

$$(i) \sum_{j \in J} y_j = 1,$$

$$(ii) \; y_j \ge 0, \; j \in J.$$

From the similar arguments used in previous case, the best response strategy of player 2 can be obtained by solving the following KKT conditions of [QP2]:

$$
\begin{cases}
0 \leq y \perp -\mu_2(x) - \dfrac{\Sigma_2(x)y \cdot c_{\alpha_2}}{\|\Sigma_2^{1/2}(x)y\|} - \lambda_3 e_n + \lambda_4 e_n \geq 0, \\[2mm]
0 \leq \lambda_3 \perp \sum_{j \in J} y_j - 1 \geq 0, \\[2mm]
0 \leq \lambda_4 \perp 1 - \sum_{j \in J} y_j \geq 0,
\end{cases} \tag{8}
$$

where $c_{\alpha_2} = F_{Z_2^N}^{-1}(1 - \alpha_2)$.

Nonlinear Complementarity Problem: By combining the KKT conditions given by (7) and (8), a Nash equilibrium (x, y) can be obtained by solving the following NCP:

$$
0 \leq \zeta \perp G(\zeta) \geq 0, \tag{9}
$$

where $\zeta, G(\zeta) \in \mathbb{R}^{m+n+4}$ are given below:

$$
\zeta^T = (x^T, y^T, \lambda_1, \lambda_2, \lambda_3, \lambda_4),
$$

$$
G(\zeta) = \begin{pmatrix}
-\mu_1(y) - \dfrac{\Sigma_1(y)x \cdot c_{\alpha_1}}{\|\Sigma_1^{1/2}(y)x\|} - \lambda_1 e_m + \lambda_2 e_m \\[2mm]
-\mu_2(x) - \dfrac{\Sigma_2(x)y \cdot c_{\alpha_2}}{\|\Sigma_2^{1/2}(x)y\|} - \lambda_3 e_n + \lambda_4 e_n \\[2mm]
\sum_{i \in I} x_i - 1 \\[1mm]
1 - \sum_{i \in I} x_i \\[1mm]
\sum_{j \in J} y_j - 1 \\[1mm]
1 - \sum_{j \in J} y_j
\end{pmatrix}.
$$

For given k, l, $\mathbf{0}_{k \times l}$ is a $k \times l$ zero matrix and $\mathbf{0}_k$ is a $k \times 1$ zero vector. Define,

$$
Q = \begin{pmatrix}
\mathbf{0}_{m \times m} & -\mu_1 & -e_m & e_m & \mathbf{0}_m & \mathbf{0}_m \\
-\mu_2^T & \mathbf{0}_{n \times n} & \mathbf{0}_n & \mathbf{0}_n & -e_n & e_n \\
e_m^T & \mathbf{0}_n^T & 0 & 0 & 0 & 0 \\
-e_m^T & \mathbf{0}_n^T & 0 & 0 & 0 & 0 \\
\mathbf{0}_m^T & e_n^T & 0 & 0 & 0 & 0 \\
\mathbf{0}_m^T & -e_n^T & 0 & 0 & 0 & 0
\end{pmatrix},
$$

where $\mu_1 = (\mu_{1,ij})_{i \in I, j \in J}$, $\mu_2 = (\mu_{2,ij})_{i \in I, j \in J}$ are $m \times n$ matrices. Define,

$$
R(\zeta) = \begin{pmatrix}
\dfrac{-c_{\alpha_1} \cdot \Sigma_1(y)}{\|\Sigma_1^{1/2}(y)x\|} & \mathbf{0}_{m \times n} & \mathbf{0}_{m \times 4} \\[2mm]
\mathbf{0}_{n \times m} & \dfrac{-c_{\alpha_2} \cdot \Sigma_2(x)}{\|\Sigma_2^{1/2}(x)y\|} & \mathbf{0}_{n \times 4} \\[2mm]
\mathbf{0}_{4 \times m} & \mathbf{0}_{4 \times n} & \mathbf{0}_{4 \times 4}
\end{pmatrix}, \quad r = \begin{pmatrix}
\mathbf{0}_m \\
\mathbf{0}_n \\
-1 \\
1 \\
-1 \\
1
\end{pmatrix}.
$$

Then, $G(\zeta) = (Q + R(\zeta))\zeta + r$.

Theorem 2. *Consider a bimatrix game (A^w, B^w). All the components of matrix A^w are independent normal random variables, where the mean and variance of a_{ij}^w, $i \in I$, $j \in J$, are $\mu_{1,ij}$ and $\sigma_{1,ij}^2$ respectively, and all the components of matrix B^w are also independent normal random variables, where the mean and variance of b_{ij}^w, $i \in I$, $j \in J$, are $\mu_{2,ij}$ and $\sigma_{2,ij}^2$ respectively. Let $\zeta^{*T} = (x^{*T}, y^{*T}, \lambda_1^*, \lambda_2^*, \lambda_3^*, \lambda_4^*)$ be a vector. Then, the strategy part (x^*, y^*) of ζ^* is a Nash equilibrium of the chance-constrained game for a given $\alpha \in [0.5, 1]^2$ if and only if ζ^* is a solution of NCP (9).*

Proof. Let $\alpha \in [0.5, 1]^2$, then (x^*, y^*) is a Nash equilibrium of the chance-constrained game if and only if x^* is an optimal solution of [QP1] for fixed y^* and y^* is an optimal solution of [QP2] for fixed x^*. Since, [QP1] and [QP2] are convex optimization problems and linear independence constraint qualification holds at all feasible points, then the KKT conditions (7) and (8) are both necessary and sufficient conditions for optimality. Then, the proof follows by combining the KKT conditions (7) and (8).

For computational purpose freely available solvers for complementarity problems can be used, e.g., see [11, 20, 26].

Special Case. Here we consider the case where the components of payoff matrices A^w and B^w are independent as well as identically distributed. We assume that the components of matrix A^w are independent and identically distributed (i.i.d.) normal random variables with mean μ_1 and variance σ_1^2, and the components of matrix B^w are i.i.d. normal random variables with mean μ_2 and variance σ_2^2.

Theorem 3. *Consider a bimatrix game (A^w, B^w) where all the components of matrix A^w are i.i.d. normal random variables with mean μ_1 and variance σ_1^2, and all the components of matrix B^w are also i.i.d. normal random variables with mean μ_2 and variance σ_2^2. The strategy pair (x^*, y^*), where,*

$$x_i^* = \frac{1}{m}, \ \forall \, i \in I, \quad y_j^* = \frac{1}{n}, \forall \, j \in J, \tag{10}$$

is a Nash equilibrium of chance-constrained game for all $\alpha \in [0.5, 1]^2$.

Proof. For all (x, y), we have $\mu_1(y) = \mu_1 e_m$ and $\mu_2(x) = \mu_2 e_n$ because $\mu_{1,ij} = \mu_1$ and $\mu_{2,ij} = \mu_2$ for all $i \in I$, $j \in J$, and $\Sigma_1(y) = \sigma_1^2 \|y\|^2 I_{m \times m}$ and $\Sigma_2(x) = \sigma_2^2 \|x\|^2 I_{n \times n}$ because $\sigma_{1,ij}^2 = \sigma_1^2$ and $\sigma_{2,ij}^2 = \sigma_2^2$ for all $i \in I, j \in J$; $I_{k \times k}$ is a $k \times k$ identity matrix. Using above expressions, and replacing the difference of two non-negative variable by a free variable, we have the following NCP equivalent to (9):

$$\begin{cases} 0 \le x \perp -\mu_1 e_m - \dfrac{\sigma_1 ||y|| x \cdot c_{\alpha_1}}{||x||} - \lambda_1 e_m \ge 0, \\[3mm] 0 \le y \perp -\mu_2 e_n - \dfrac{\sigma_2 ||x|| y \cdot c_{\alpha_2}}{||y||} - \lambda_2 e_n \ge 0, \\[3mm] \displaystyle\sum_{i \in I} x_i = 1, \ \sum_{j \in J} y_j = 1 \\[3mm] \lambda_1, \ \lambda_2 \in \mathbb{R}. \end{cases} \tag{11}$$

Consider the Lagrange multipliers $(\lambda_1^*, \lambda_2^*)$ as follows:

$$\lambda_1^* = -\frac{\sigma_1 \cdot c_{\alpha_1}}{\sqrt{mn}} - \mu_1, \ \lambda_2^* = -\frac{\sigma_2 \cdot c_{\alpha_2}}{\sqrt{mn}} - \mu_2.$$

It is easy to check that $(x^*, y^*, \lambda_1^*, \lambda_2^*)$ is a solution of NCP (11). That is, (x^*, y^*) defined by (10) is a Nash equilibrium of chance-constrained game.

3.2   Payoffs Following Cauchy Distribution

We assume that all the components of matrix A^w are independent Cauchy random variables, where the location and scale parameters of a_{ij}^w, $i \in I$, $j \in J$, are $\mu_{1,ij}$ and $\sigma_{1,ij}$ respectively, and all the components of matrix B^w are independent Cauchy random variables, where the location and scale parameters of b_{ij}^w, $i \in I$, $j \in J$, are $\mu_{2,ij}$ and $\sigma_{2,ij}$ respectively. Since, a linear combination of the independent Cauchy random variables is a Cauchy random variable [13], then, for a given strategy pair (x, y), the payoff $x^T A^w y$ of player 1 follows a Cauchy distribution with location parameter $\mu_1(x, y) = \sum_{i \in I, j \in J} x_i y_j \mu_{1,ij}$ and scale parameter $\sigma_1(x, y) = \sum_{i \in I, j \in J} x_i y_j \sigma_{1,ij}$, and the payoff $x^T B^w y$ of player 2 follows a Cauchy distribution with location parameter $\mu_2(x, y) = \sum_{i \in I, j \in J} x_i y_j \mu_{2,ij}$ and scale parameter $\sigma_2(x, y) = \sum_{i \in I, j \in J} x_i y_j \sigma_{2,ij}$. Then, $Z_1^C = \frac{x^T A^w y - \mu_1(x,y)}{\sigma_1(x,y)}$ and $Z_2^C = \frac{x^T B^w y - \mu_2(x,y)}{\sigma_2(x,y)}$ follow a standard Cauchy distribution. Let $F_{Z_1^C}^{-1}(\cdot)$ and $F_{Z_2^C}^{-1}(\cdot)$ be the quantile functions of a standard Cauchy distribution. For more details about Cauchy distribution see [13]. Similar to the normal distribution case, for a given strategy pair (x, y) and a given α the payoff of player 1 is given by

$$u_1^{\alpha_1}(x, y) = \sup \left\{ u \Big| F_{Z_1^C} \left(\frac{u - \mu_1(x, y)}{\sigma_1(x, y)} \right) \le 1 - \alpha_1 \right\}$$

$$= \sup \left\{ u | u \le \mu_1(x, y) + \sigma_1(x, y) F_{Z_1^C}^{-1}(1 - \alpha_1) \right\}.$$

That is,

$$u_1^{\alpha_1}(x, y) = \sum_{i \in I, j \in J} x_i y_j \left(\mu_{1,ij} + \sigma_{1,ij} F_{Z_1^C}^{-1}(1 - \alpha_1) \right). \tag{12}$$

Similarly, the payoff of player 2 is given by

$$u_2^{\alpha_2}(x,y) = \sum_{i \in I, j \in J} x_i y_j \left(\mu_{2,ij} + \sigma_{2,ij} F_{Z_2^C}^{-1}(1 - \alpha_2) \right). \tag{13}$$

The quantile function of a standard Cauchy distribution is not finite at 0 and 1. Therefore, we consider the case of $\alpha \in (0,1)^2$, so that payoff functions defined by (12) and (13) have finite values. Define, a matrix $\tilde{A}(\alpha_1) = (\tilde{a}_{ij}(\alpha_1))_{i \in I, j \in J}$, where

$$\tilde{a}_{ij}(\alpha_1) = \mu_{1,ij} + \sigma_{1,ij} F_{Z_1^C}^{-1}(1 - \alpha_1), \tag{14}$$

and a matrix $\tilde{B}(\alpha_2) = \left(\tilde{b}_{ij}(\alpha_2) \right)_{i \in I, j \in J}$, where

$$\tilde{b}_{ij}(\alpha_2) = \mu_{2,ij} + \sigma_{2,ij} F_{Z_2^C}^{-1}(1 - \alpha_2). \tag{15}$$

Then, we can write (12) as

$$u_1^{\alpha_1}(x,y) = x^T \tilde{A}(\alpha_1) y,$$

and we can write (13) as

$$u_2^{\alpha_2}(x,y) = x^T \tilde{B}(\alpha_2) y.$$

Then, for a given $\alpha \in (0,1)^2$, the chance-constrained game is equivalent to the bimatrix game $(\tilde{A}(\alpha_1), \tilde{B}(\alpha_2))$.

Theorem 4. *Consider a bimatrix game (A^w, B^w). If all the components of matrix A^w are independent Cauchy random variables, and all the components of matrix B^w are also independent Cauchy random variables, there exists a mixed strategy Nash equilibrium for a chance-constrained game for all $\alpha \in (0,1)^2$.*

Proof. For each $\alpha \in (0,1)^2$ the chance-constrained game is equivalent to the bimatrix game $(\tilde{A}(\alpha_1), \tilde{B}(\alpha_2))$. Hence, the existence of a Nash equilibrium in mixed strategies follows from [21].

Remark 1. For case of i.i.d. Cauchy random variables each strategy pair (x,y) is a Nash equilibrium because from (12) and (13) the payoff functions of both the players are constant.

Linear Complementarity Problem. For a given matrix $N = [N_{ij}]$, $N > 0$ means that $N_{ij} > 0$ for all i, j. Let E be the $m \times n$ matrix with all 1's. Let k be the large enough such that $kE^T - (\tilde{B}(\alpha_2))^T > 0$ and $kE - \tilde{A}(\alpha_1) > 0$. Then, from [15,16] it follows that for a given α, a Nash equilibrium of the chance-constrained game can be obtained by following LCP:

$$0 \le z \perp Mz + q \ge 0, \tag{16}$$

where

$$z = \begin{pmatrix} x \\ y \end{pmatrix}, \quad M = \begin{pmatrix} \mathbf{0}_{m \times m} & kE - \tilde{A}(\alpha_1) \\ kE^T - (\tilde{B}(\alpha_2))^T & \mathbf{0}_{n \times n} \end{pmatrix},$$

$$q = \begin{pmatrix} -e_m \\ -e_n \end{pmatrix}.$$

Theorem 5. *Consider a bimatrix game* (A^w, B^w) *where all the components of matrix* A^w *are independent Cauchy random variables, and all the components of matrix* B^w *are also independent Cauchy random variables, then,*

1. *For a* $\alpha \in (0,1)^2$, *if* (x^*, y^*) *is a Nash equilibrium of the chance-constrained game,* $z^{*T} = \left(\frac{x^{*T}}{k - x^{*T} \tilde{B}(\alpha_2) y^*}, \frac{y^{*T}}{k - x^{*T} \tilde{A}(\alpha_1) y^*} \right)$ *is a solution of LCP (16) at* α.
2. *For a* $\alpha \in (0,1)^2$, *if* $\bar{z}^T = (\bar{x}^T, \bar{y}^T)$ *is a solution of LCP (16),* $(x^*, y^*) = \left(\frac{\bar{x}}{\sum_{i \in I} \bar{x}_i}, \frac{\bar{y}}{\sum_{j \in J} \bar{y}_j} \right)$ *is a Nash equilibrium of the chance-constrained game at* α.

Proof. For a $\alpha \in (0,1)^2$, the chance-constrained game corresponding to Cauchy distribution is equivalent to a bimatrix game $(\tilde{A}(\alpha_1), \tilde{B}(\alpha_2))$, where $\tilde{A}(\alpha_1)$ and $\tilde{B}(\alpha_2)$ is defined by (14) and (15) respectively. Then, the proof follows from [16]. $\quad\blacksquare$

Numerical Results. We consider few instances of random bimatrix game of different sizes. We compute the Nash equilibria of corresponding chance-constrained game by using Lemke-Howson algorithm [16]. We use the MATLAB code of Lemke-Howson algorithm given in [14].

(i) 3×3 *random bimatrix game:* We consider five instances of random bimatrix game of size 3×3. The datasets consisting of location parameters $\mu_1 = [\mu_{1,ij}]$, $\mu_2 = [\mu_{2,ij}]$, and scale parameters $\sigma_1 = [\sigma_{1,ij}]$, $\sigma_2 = [\sigma_{2,ij}]$ of independent Cauchy random variables that characterize chance-constrained game are follows:

1. $\mu_1 = \begin{pmatrix} 1 & 2 & 1 \\ 2 & 3 & 1 \\ 1 & 2 & 3 \end{pmatrix}$, $\sigma_1 = \begin{pmatrix} 1 & 1 & 2 \\ 1 & 2 & 3 \\ 2 & 1 & 2 \end{pmatrix}$, $\mu_2 = \begin{pmatrix} 2 & 1 & 2 \\ 3 & 2 & 1 \\ 1 & 2 & 3 \end{pmatrix}$, $\sigma_2 = \begin{pmatrix} 1 & 2 & 3 \\ 3 & 1 & 2 \\ 2 & 3 & 1 \end{pmatrix}$.

2. $\mu_1 = \begin{pmatrix} 1 & 1 & 2 \\ 2 & 1 & 1 \\ 2 & 1 & 3 \end{pmatrix}$, $\sigma_1 = \begin{pmatrix} 2 & 2 & 3 \\ 3 & 2 & 1 \\ 1 & 2 & 3 \end{pmatrix}$, $\mu_2 = \begin{pmatrix} 2 & 2 & 1 \\ 3 & 2 & 3 \\ 2 & 1 & 2 \end{pmatrix}$, $\sigma_2 = \begin{pmatrix} 1 & 2 & 2 \\ 2 & 3 & 1 \\ 2 & 1 & 3 \end{pmatrix}$.

3. $\mu_1 = \begin{pmatrix} 2 & 1 & 3 \\ 3 & 2 & 1 \\ 1 & 3 & 2 \end{pmatrix}$, $\sigma_1 = \begin{pmatrix} 2 & 3 & 1 \\ 3 & 1 & 2 \\ 1 & 2 & 3 \end{pmatrix}$, $\mu_2 = \begin{pmatrix} 1 & 2 & 3 \\ 2 & 1 & 3 \\ 3 & 1 & 2 \end{pmatrix}$, $\sigma_2 = \begin{pmatrix} 1 & 1 & 2 \\ 1 & 2 & 1 \\ 3 & 1 & 1 \end{pmatrix}$.

4. $\mu_1 = \begin{pmatrix} 3 & 1 & 2 \\ 2 & 1 & 3 \\ 1 & 2 & 3 \end{pmatrix}$, $\sigma_1 = \begin{pmatrix} 2 & 4 & 1 \\ 1 & 2 & 3 \\ 3 & 2 & 1 \end{pmatrix}$, $\mu_2 = \begin{pmatrix} 4 & 1 & 3 \\ 3 & 2 & 4 \\ 2 & 1 & 3 \end{pmatrix}$, $\sigma_2 = \begin{pmatrix} 5 & 2 & 3 \\ 3 & 2 & 1 \\ 4 & 2 & 3 \end{pmatrix}$.

5. $\mu_1 = \begin{pmatrix} 1 & 2 & 1 \\ 2 & 3 & 1 \\ 1 & 2 & 3 \end{pmatrix}$, $\sigma_1 = \begin{pmatrix} 2 & 2 & 3 \\ 3 & 2 & 1 \\ 1 & 2 & 3 \end{pmatrix}$, $\mu_2 = \begin{pmatrix} 1 & 2 & 3 \\ 2 & 1 & 3 \\ 3 & 1 & 2 \end{pmatrix}$, $\sigma_2 = \begin{pmatrix} 5 & 2 & 3 \\ 3 & 2 & 1 \\ 4 & 2 & 3 \end{pmatrix}$.

The entries of μ_1, σ_1, μ_2, σ_2 defined above are the location and scale parameters of corresponding independent Cauchy random variables. For example, in dataset 1 random payoff a_{11} is a Cauchy random variable with location parameter 1 and scale parameter 1. Table 1 summarizes the Nash equilibria of chance-constrained game corresponding to datasets given for five instances of 3×3 random bimatrix game.

Table 1. Nash equilibria for various values of α.

No.	α		Nash equilibrium	
	α_1	α_2	x^*	y^*
1	0.4	0.4	$(0, 0, 1)$	$(0, 0, 1)$
	0.5	0.5	$(0, 1, 0)$	$(1, 0, 0)$
	0.7	0.7	$(0, 1, 0)$	$(0, 1, 0)$
2	0.4	0.4	$(1, 0, 0)$	$(0, 1, 0)$
	0.5	0.5	$(0, 1, 0)$	$(1, 0, 0)$
	0.7	0.7	$(0, 0, 1)$	$(1, 0, 0)$
3	0.4	0.4	$(1, 0, 0)$	$(0, 0, 1)$
	0.5	0.5	$(1, 0, 0)$	$(0, 0, 1)$
	0.7	0.7	$(1, 0, 0)$	$(0, 0, 1)$
4	0.4	0.4	$(1, 0, 0)$	$(1, 0, 0)$
	0.5	0.5	$(1, 0, 0)$	$(1, 0, 0)$
	0.7	0.7	$(0, 0, 1)$	$(0, 0, 1)$
5	0.4	0.4	$\left(0, \frac{791}{1000}, \frac{209}{1000}\right)$	$\left(\frac{616}{1000}, 0, \frac{384}{1000}\right)$
	0.5	0.5	$\left(0, \frac{1}{2}, \frac{1}{2}\right)$	$\left(\frac{2}{3}, 0, \frac{1}{3}\right)$
	0.7	0.7	$(0, 0, 1)$	$(1, 0, 0)$

(*ii*) 5×5 *random bimatrix game:* We consider two instances of random bimatrix game of size 5×5. The location parameters μ_1, μ_2, and scale parameters σ_1, σ_2 of independent Cauchy random variables are as follows:

$$1. \ \mu_1 = \begin{pmatrix} 1 & 2 & 1 & 1 & 3 \\ 2 & 3 & 1 & 1 & 2 \\ 1 & 2 & 3 & 2 & 3 \\ 2 & 1 & 3 & 4 & 2 \\ 1 & 2 & 4 & 5 & 2 \end{pmatrix}, \sigma_1 = \begin{pmatrix} 2 & 2 & 3 & 2 & 1 \\ 1 & 2 & 3 & 2 & 1 \\ 1 & 2 & 3 & 3 & 1 \\ 2 & 1 & 3 & 4 & 2 \\ 3 & 1 & 2 & 5 & 2 \end{pmatrix},$$

$$\mu_2 = \begin{pmatrix} 1 & 2 & 3 & 2 & 1 \\ 3 & 2 & 2 & 1 & 3 \\ 1 & 2 & 3 & 1 & 2 \\ 2 & 1 & 4 & 2 & 1 \\ 1 & 1 & 2 & 1 & 3 \end{pmatrix}, \sigma_2 = \begin{pmatrix} 5 & 2 & 3 & 2 & 3 \\ 2 & 4 & 3 & 2 & 1 \\ 1 & 3 & 4 & 2 & 3 \\ 2 & 1 & 3 & 5 & 1 \\ 2 & 1 & 2 & 3 & 4 \end{pmatrix}.$$

$$2.\ \mu_1 = \begin{pmatrix} 1\,2\,2\,4\,3 \\ 2\,1\,3\,2\,2 \\ 1\,2\,4\,2\,1 \\ 2\,2\,3\,4\,1 \\ 1\,2\,4\,5\,2 \end{pmatrix}, \sigma_1 = \begin{pmatrix} 2\,3\,1\,2\,1 \\ 1\,1\,3\,1\,2 \\ 3\,1\,3\,3\,1 \\ 2\,2\,5\,4\,2 \\ 3\,1\,3\,5\,2 \end{pmatrix},$$

$$\mu_2 = \begin{pmatrix} 1\,2\,3\,2\,1 \\ 3\,1\,2\,1\,4 \\ 2\,1\,3\,4\,2 \\ 3\,2\,4\,2\,1 \\ 2\,4\,2\,1\,3 \end{pmatrix}, \sigma_2 = \begin{pmatrix} 5\,2\,4\,2\,1 \\ 2\,4\,3\,2\,1 \\ 4\,3\,3\,2\,3 \\ 2\,1\,3\,5\,3 \\ 1\,3\,4\,3\,4 \end{pmatrix}.$$

Table 2 summarizes the Nash equilibria of chance-constrained game corresponding to datasets given for two instances of 5×5 random bimatrix game.

Table 2. Nash equilibria for various values of α.

No.	α		Nash Equilibrium	
	α_1	α_2	x^*	y^*
1	0.4	0.4	$(0,0,\frac{555}{1000},0,\frac{445}{1000})$	$(0,0,\frac{1}{2},0,\frac{1}{2})$
	0.5	0.5	$(0,0,\frac{1}{2},0,\frac{1}{2})$	$(0,0,\frac{1}{2},0,\frac{1}{2})$
	0.7	0.7	$(0,0,0,0,1)$	$(0,0,1,0,0)$
2	0.4	0.4	$(0,0,\frac{663}{1000},0,\frac{337}{1000})$	$(0,0,1,0,0)$
	0.5	0.5	$(0,0,\frac{1}{2},0,\frac{1}{2})$	$(0,1,0,0,0)$
	0.7	0.7	$(0,0,\frac{446}{1000},0,\frac{554}{1000})$	$(0,1,0,0,0)$

(iii) 7×7 *random bimatrix game:* We consider two instances of random bimatrix game of size 7×7. The location parameters μ_1, μ_2, and scale parameters σ_1, σ_2 of independent Cauchy random variables are as follows:

$$1.\ \mu_1 = \begin{pmatrix} 1\,2\,2\,4\,3\,2\,1 \\ 1\,1\,2\,1\,3\,2\,2 \\ 3\,2\,1\,2\,4\,2\,1 \\ 2\,4\,2\,2\,3\,4\,1 \\ 1\,2\,4\,5\,2\,2\,3 \\ 1\,3\,4\,3\,2\,2\,3 \\ 2\,1\,4\,2\,3\,2\,1 \end{pmatrix}, \sigma_1 = \begin{pmatrix} 2\,3\,1\,2\,1\,1\,2 \\ 1\,1\,3\,1\,2\,2\,4 \\ 2\,1\,3\,1\,3\,3\,1 \\ 2\,2\,5\,4\,2\,1\,3 \\ 2\,1\,3\,1\,3\,5\,2 \\ 1\,2\,3\,1\,2\,3\,2 \\ 2\,1\,4\,2\,3\,1\,2 \end{pmatrix},$$

$$\mu_2 = \begin{pmatrix} 1\,2\,3\,2\,3\,2\,1 \\ 1\,2\,3\,1\,2\,1\,4 \\ 2\,1\,2\,1\,3\,4\,2 \\ 1\,2\,3\,2\,4\,2\,1 \\ 2\,3\,1\,4\,2\,1\,3 \\ 1\,2\,3\,2\,1\,3\,4 \\ 2\,3\,1\,2\,3\,4\,2 \end{pmatrix}, \sigma_2 = \begin{pmatrix} 5\,2\,4\,2\,1\,2\,3 \\ 1\,2\,2\,4\,3\,2\,1 \\ 2\,3\,4\,3\,3\,2\,3 \\ 2\,3\,2\,1\,3\,5\,3 \\ 2\,1\,2\,3\,4\,3\,4 \\ 1\,2\,2\,3\,1\,3\,1 \\ 2\,4\,1\,2\,3\,1\,2 \end{pmatrix}.$$

$$
2.\ \mu_1 = \begin{pmatrix} 1\,2\,3\,1\,3\,4\,1 \\ 2\,1\,2\,1\,2\,4\,2 \\ 1\,2\,1\,5\,3\,2\,1 \\ 1\,3\,2\,2\,3\,2\,1 \\ 2\,3\,4\,5\,2\,1\,3 \\ 1\,3\,2\,1\,2\,4\,3 \\ 2\,1\,3\,2\,1\,2\,1 \end{pmatrix},\ \sigma_1 = \begin{pmatrix} 1\,3\,1\,2\,1\,2\,2 \\ 2\,1\,3\,1\,2\,2\,4 \\ 2\,1\,3\,2\,3\,4\,1 \\ 2\,2\,3\,4\,2\,1\,3 \\ 2\,4\,3\,1\,3\,2\,2 \\ 1\,2\,3\,2\,2\,4\,2 \\ 2\,3\,4\,1\,3\,1\,2 \end{pmatrix},
$$

$$
\mu_2 = \begin{pmatrix} 2\,1\,3\,4\,3\,2\,1 \\ 1\,2\,3\,3\,2\,1\,4 \\ 2\,1\,2\,1\,3\,4\,2 \\ 1\,2\,3\,2\,4\,2\,1 \\ 2\,3\,2\,4\,2\,1\,3 \\ 1\,2\,1\,2\,5\,3\,4 \\ 2\,3\,1\,2\,1\,4\,2 \end{pmatrix},\ \sigma_2 = \begin{pmatrix} 5\,2\,4\,3\,1\,2\,3 \\ 1\,2\,3\,4\,3\,2\,3 \\ 1\,3\,4\,2\,1\,2\,3 \\ 2\,3\,2\,2\,3\,4\,3 \\ 2\,1\,2\,2\,4\,1\,4 \\ 2\,3\,2\,3\,4\,3\,1 \\ 2\,4\,3\,2\,3\,1\,2 \end{pmatrix}.
$$

Table 3 summarizes the Nash equilibria of chance-constrained game corresponding to datasets given for two instances of 7×7 random bimatrix game.

Table 3. Nash equilibria for various values of α.

No.	α		Nash Equilibrium	
	α_1	α_2	x^*	y^*
1	0.4	0.4	$\left(0,0,\frac{2}{3},\frac{1}{3},0,0,0\right)$	$\left(0,0,0,0,\frac{505}{1000},\frac{495}{1000},0\right)$
	0.5	0.5	$\left(0,0,\frac{2}{3},\frac{1}{3},0,0,0\right)$	$\left(0,0,0,0,\frac{2}{3},\frac{1}{3},0\right)$
	0.7	0.7	$(1,0,0,0,0,0,0)$	$(0,0,0,0,1,0,0)$
2	0.4	0.4	$\left(\frac{1}{5},0,\frac{13}{25},0,\frac{7}{25},0,0\right)$	$\left(0,0,\frac{13}{50},0,\frac{675}{1000},\frac{65}{1000},0\right)$
	0.5	0.5	$\left(\frac{1}{2},0,\frac{1}{2},0,0,0,0\right)$	$(0,0,0,0,1,0,0)$
	0.7	0.7	$(1,0,0,0,0,0,0)$	$(0,0,0,0,1,0,0)$

4 Conclusions

We formulate a random bimatrix game, as a chance-constrained game by defining the payoff of each player using a chance constraint. We show a one-to-one correspondence between a Nash equilibrium problem, corresponding to independent normally distributed payoffs, and a certain nonlinear complementarity problem. Further if the payoffs are also identically distributed, a uniform strategy pair is a Nash equilibrium. We show that a Nash equilibrium problem corresponding to independent Cauchy distributed payoffs is equivalent to a linear complementarity problem. Further if the payoffs are also identically distributed, each strategy pair would be a Nash equilibrium. For chance-constrained game corresponding to Cauchy distribution, we carried out numerical experiments by considering random instances of various sizes. We use Lemke-Howson algorithm to compute the Nash equilibria.

Recently, the electricity markets over the past few years have been transformed from nationalized monopolies into competitive markets with privately owned participants. The uncertainties in electricity markets are present due to various external factors, e.g., random demand, wind generation. The chance-constrained game formulation proposed in the paper would be useful when the electricity firms are risk averse. For example, it can be used to model the Cournot or Bertrand competition among electricity firms where the action sets of the firms are finite (which is possible using discretization), and the demand or cost functions are uncertain. The computational approaches developed in this paper can be applied to compute the Nash equilibria of these games.

Acknowledgements. This research was supported by Fondation DIGITEO, SUN grant No. 2014-0822D.

References

1. Bazaraa, M., Sherali, H., Shetty, C.: Nonlinear Programming Theory and Algorithms, 3rd edn. Wiley, New York (2006)
2. Blau, R.A.: Random-payoff two person zero-sum games. Oper. Res. **22**(6), 1243–1251 (1974)
3. Cassidy, R.G., Field, C.A., Kirby, M.J.L.: Solution of a satisficing model for random payoff games. Manage. Sci. **19**(3), 266–271 (1972)
4. Charnes, A., Cooper, W.W.: Deterministic equivalents for optimizing and satisficing under chance constraints. Oper. Res. **11**(1), 18–39 (1963)
5. Charnes, A., Kirby, M.J.L., Raike, W.M.: Zero-zero chance-constrained games. Theory Probab. Appl. **13**(4), 628–646 (1968)
6. Cheng, J., Lisser, A.: A second-order cone programming approach for linear programs with joint probabilistic constraints. Oper. Res. Lett. **40**(5), 325–328 (2012)
7. Collins, W.D., Hu, C.: Studying interval valued matrix games with fuzzy logic. Soft. Comput. **12**, 147–155 (2008)
8. Couchman, P., Kouvaritakis, B., Cannon, M., Prashad, F.: Gaming strategy for electric power with random demand. IEEE Trans. Power Syst. **20**(3), 1283–1292 (2005)
9. DeMiguel, V., Xu, H.: A stochastic multiple leader stackelberg model: analysis, computation, and application. Oper. Res. **57**(5), 1220–1235 (2009)
10. Li, D.-F., Nan, J.-X., Zhang, M.J.: Interval programming models for matrix games with interval payoffs. Optim. Meth. Softw. **27**(1), 1–16 (2012)
11. Ferris, M.C., Munson, T.S.: Complementarity problems in GAMS and the PATH solver. J. Econ. Dyn. Control **24**, 165–188 (2000)
12. Jadamba, B., Raciti, F.: Variational inequality approach to stochastic Nash equilibrium problems with an application to cournot oligopoly. J. Optim. Theor. Appl. **165**(3), 1050–1070 (2015)
13. Johnson, N.L., Kotz, S., Balakrishnan, N.: Continuous Univariate Distributions, vol. 1, 2nd edn. Wiley, New York (1994)
14. Katzwer, R.: Lemke-Howson Algorithm for 2-Player Games (2013). File ID: #44279 Version: 1.3
15. Lemke, C.E.: Bimatrix equilibrium points and mathematical programming. Manage. Sci. **11**(7), 681–689 (1965)

16. Lemke, C., Howson, J.: Equilibrium points of bimatrix games. SIAM J. **12**, 413–423 (1964)
17. Li, D.F.: Linear programming approach to solve interval-valued matrix games. J. Omega **39**(6), 655–666 (2011)
18. Mazadi, M., Rosehart, W.D., Zareipour, H., Malik, O.P., Oloomi, M.: Impact of wind integration on electricity markets: a chance-constrained Nash Cournot model. Int. Trans. Electr. Energy Syst. **23**(1), 83–96 (2013)
19. Mitchell, C., Hu, C., Chen, B., Nooner, M., Young, P.: A computational study of interval-valued matrix games. In: International Conference on Computational Science and Computational Intelligence (2014)
20. Munson, T.S.: Algorithms and environments for complementarity. Ph.D. thesis, University of Wisconsin, Madison (2000)
21. Nash, J.F.: Equilibrium points in n-person games. Proc. Natl. Acad. Sci. **36**(1), 48–49 (1950)
22. Neumann, J.V.: Zur theorie der gesellschaftsspiele. Math. Annalen **100**(1), 295–320 (1928)
23. Nocedal, J., Wright, S.J.: Numerical Optimization, 2nd edn. Springer, New York (2006)
24. Prékopa, A.: Stochastic Programming. Springer, Netherlands (1995)
25. Ravat, U., Shanbhag, U.V.: On The characterization of solution sets of smooth and nonsmooth convex stochastic Nash games. SIAM J. Optim. **21**(3), 1168–1199 (2011)
26. Schmelzer, S.: COMPASS: a free solver for mixed complementarity problems. Master's thesis, Universität Wien (2012)
27. Singh, V.V., Jouini, O., Lisser, A.: Existence of Nash equilibrium for chance-constrained games. Oper. Res. Lett. **44**(5), 640–644 (2016)
28. Singh, V.V., Jouini, O., Lisser, A.: Distributionally robust chance-constrained games: existence and characterization of Nash equilibrium. Optim. Lett. (2016). doi:10.1007/s11590-016-1077-6
29. Song, T.: On random payoff matrix games. In: Phillips, F.Y., Rousseau, J.J. (eds.) Systems and Management Science by Extremal Methods, pp. 291–308. Springer, Heidelberg (1992)
30. Valenzuela, J., Mazumdar, M.: Cournot prices considering generator availability and demand uncertainty. IEEE Trans. Power Syst. **22**(1), 116–125 (2007)
31. Wolf, D.D., Smeers, Y.: A stochastic version of a Stackelberg-Nash-Cournot equilibrium model. Manage. Sci. **43**(2), 190–197 (1997)
32. Xu, H., Zhang, D.: Stochastic Nash equilibrium problems: sample average approximation and applications. Comput. Optim. Appl. **55**(3), 597–645 (2013)

Applications

Distributed Patrolling with Two-Speed Robots (and an Application to Transportation)

Jurek Czyzowicz[1], Konstantinos Georgiou[2(✉)], Evangelos Kranakis[3], Fraser MacQuarrie[3], and Dominik Pajak[4]

[1] Departemant d'Informatique, Universite du Quebec en Outaouais, Gatineau, QC, Canada
[2] Department of Mathematics, Ryerson University, Toronto, ON, Canada
konstantinos@ryerson.ca
[3] School of Computer Science, Carleton University, Ottawa, ON, Canada
[4] Department of Computer Science FFPT, Wroclaw University of Technology, Wroclaw, Poland

Abstract. We initiate the study of patrolling a unit interval with primitive two-speed, autonomous robots, i.e. robots without memory, no communication capabilities and no computation power. Robots have only two moving-states, one for *patrolling* and one for *walking*, each associated with a direction and speed. The robots are moving perpetually, and their moving-states and moving directions change only when they collide. Such a dynamic system induces the so-called *idleness* for patrolling a unit interval, i.e. the smallest time interval within which every point of the domain is patrolled by some robot. Our main technical contribution is an analytic study of the induced dynamic system of robots, which allows us to decide efficiently whether or not the system converges to a stable configuration that is also shown to be optimal.

As a warm-up for our main result, we show how robots can be centrally coordinated, carefully choosing initial locations, so that the induced idleness is optimal. Our main result pertaining to the idleness of primitive robots follows by a technical analysis of their collision locations, which we show, under some conditions, converge to the optimal initial locations for non-distributed robots.

Our result finds an application to a transportation problem concerning *Scheduling with Regular Delivery*. In this optimization problem, an infinite quantity of a commodity, residing at an endpoint of an interval, needs to be transported to the other endpoint. To that end, we show that the already established patrolling schedules of an interval correspond to optimal strategies that guarantee that the flow of the commodity is the largest possible.

Keywords: Mobile robots · Patrolling · Idleness · Speed · Walking · Scheduling

J. Czyzowicz and E. Kranakis—Research supported in part by NSERC Discovery grant.

© Springer International Publishing AG 2017
B. Vitoriano and G.H. Parlier (Eds.): ICORES 2016, CCIS 695, pp. 71–95, 2017.
DOI: 10.1007/978-3-319-53982-9_5

1   Introduction

Fence patrolling is concerned with perpetual monitoring of a domain, modelled as a unit segment, by a set of mobile robots. As the robots cannot continuously monitor all points of the domain, a standard efficiency measure of patrolling is the notion of *idleness* - the size of the minimal time interval for which all points of the domain are always visited (independently of the start of this interval). Patrolling algorithms attempt to produce the schedules (i.e. the trajectories of the robots in time) minimizing idleness.

In previous research, the robots were supposed to have either the same or possibly different maximal speeds of their patrolling movements. In this paper we assume that the patrolling activity requires more elaborate work than just walking within the domain, e.g., when *foraging* or *harvesting* which takes longer than walking, and in computing applications, *web page indexing, forensic search, code inspection, packet sniffing* which require a more involved inspection. Consequently, we assume the maximal patrolling speeds to be strictly smaller than the maximal walking speeds.

We suppose that each robot is capable of patrolling in only one direction (arbitrarily chosen) of the segment while walking is permitted in both directions. It is worth noting that the fence problem for robots patrolling in both directions appears to be quite difficult.

In fact, even when the walking and patrolling speeds are assumed to be the same (though perhaps distinct for different robots) the optimal schedules for fence patrolling are known for up to three robots only (cf. [15, 19, 26]). In this case the *partition strategy*, when each robot operates within a different subsegment of the domain, has been proven to work. On the other hand, this strategy has been shown insufficient for larger collections of robots and no solution is known in the general case. In our paper we suppose that each robot may patrol in only one direction of the segment (arbitrarily chosen) while walking is permitted in both directions. Such one-way patrolling assumption is quite natural for humans (e.g. text inspection). (Anyone who played the backward spelling game must agree.) In the case of physical devices, when heavy goods are transported along a road (or up the hill) the speeds with or without loads, up or down the hill are not the same.

This one-way patrolling assumption allowed us to solve optimally the patrolling problem in the centralized scenario. We also show how to apply our algorithm to natural optimization problem in transportation concerning *Scheduling with Regular Delivery*: Suppose that a commodity produced at point A has to be delivered along some path to point B at regular time intervals. Consider also a collection of diverse transporters whose speeds during transportation are smaller than their speeds while traveling empty. We are interested for the shortest time (interval) T, such that an amount of the commodity is delivered within each time interval $[t, t + T)$.

Next, and more surprisingly, using a very weak collection of mobile robots with only primitive capabilities, we are able to design a distributed strategy converging to the same optimal solution thus achieving the same idleness as the optimal centralized algorithm. Our robots are anonymous, oblivious, silent (no communication permitted) and they cannot process any information, as well as they are not aware of their numbers or their speeds. Our robots are also subject to *passive mobility*, i.e. their movement is dynamically controlled by their contacts with the environment. More specifically, robots are only capable of walking or patrolling with maximal speed and their perception mechanism permits only to recognize a collision with obstacles (i.e. the endpoints of the segment or another robot). This is the first study of the patrolling problem in a decentralized setting thus leading to the design of a distributed, self-stabilizing algorithm (cf [18] [EWD386, pp. 34–35], which discusses a solution to a cyclic relaxation problem).

1.1 Related Work

Patrolling consists of moving perpetually around an area in order to survey, protect or supervise it, and it has been studied intensively in robotics [3,12,20,21,25,28,32] where it is often viewed as a version of terrain *coverage*, a central task in robotics. It is useful in ecological monitoring, detecting intrusion, monitoring and locating objects or humans (that may need to be rescued from a disaster), detecting network failures or even discovering web pages which need to be indexed by search engines [28]. Boundary and area patrolling have been studied in [2,20,21,24,31] with ad hoc approaches emphasizing experimental results (e.g. [28]), uncertainty of the model and robustness of the solutions when failures are possible (e.g., [20,21,25]) or non-deterministic solutions (e.g., [2]). In the experimental paper [28] several fundamental theoretical concepts can be found which are related to patrolling, including models of robots (e.g., visibility or depth of perception), means of communication or motion coordination, as well as measures of algorithmic efficiency.

Idleness is the accepted measure of algorithmic efficiency of patrolling and is related to the frequency with which the points of the environment are visited [3,5,12,20,21,28] (this last criterion was first introduced in [28]). Depending on the requirements, idleness may sometimes be viewed as the average [20], worst-case [9,32], probabilistic [2] or experimentally verified [28] time elapsed since the last visit of a node [3,12]. Terms like *blanket time* [32] or *refresh time* [31] have also been used in the literature. Diverse approaches to patrolling based on idleness criteria are discussed in [3]. Patrolling as a game between patrollers and intruder is studied in [4–6]. Elsewhere, patrolling is studied based on swarm or ant-based algorithms [22,30,32]. Robots are memoryless (or having small memory), decentralized [30] with no explicit communication permitted either with other robots or the central station, and may have local sensing capabilities [22]. Ant-like algorithms usually mark the visited nodes of the graph and [32] presents an evolutionary process. This paper shows that a team of memoryless robots, by leaving marks at the nodes while walking through them, after relatively short time stabilizes to the patrolling scheme in which the frequency of the traversed edges is uniform to a factor of two (i.e., the number of traversals of the most often visited edge is at most twice the number of traversal of the least visited one).

Theoretical graph-based approaches to patrolling can be found in [12]. The two basic methods are referred to as *cyclic strategies*, where a single cycle spanning the entire graph is constructed with the robots assigned to consecutively traverse this cycle in the same direction, and as *partition-based strategies*, where the region is split into a number of either disjoint or overlapping portions to be patrolled by subsets of robots assigned to these regions. The environment and the time considered in the models studied are usually discrete in an underlying graph environment. When the environment is geometric, a *skeletonization* technique may be applied, with the terrain first partitioned into cells, and then graph-theoretic methods are used. Cyclic strategies usually rely either on TSP-related solutions or spanning tree-based approaches. For example, spanning tree coverage, a technique first introduced in [23], was later extended and used in [1,20,25]. This technique is a version of the skeletonization approach where the two-dimensional grid approximating the terrain is constructed and a Hamiltonian path present in the grid is used for patrolling. In [31], polynomial-time patrolling solutions for lines and trees are proposed. For the case of cyclic graphs, [31] proves the NP-hardness of the problem and a constant-factor approximation is proposed.

Optimal patrolling with same-speed robots in mixed domains, where regions to be traversed are fragmented by components that do not need to be monitored, is studied in [14]. Patrolling with robots that do not necessarily have identical speeds offers several surprises both in terms of the difficulty of the problem as well as in terms of the algorithmic results obtained. Such a study has been initiated in [15] and investigated further in [19,26]. The partition strategy, where each robot patrols and walks along a separate area, has been proven to work for two robots in [15], and for three in [26].

Standard capabilities of mobile robots usually include communication, computation, and environment perception. For many reasons (e.g. production cost, limited or specific applications) one may wish to deal with robots of reduced ability, especially if they are needed in large numbers. In such cases, feasibility issues, rather than computation efficiency are sought [7,8,10,13]. [7] introduced *population protocols* (see also [8,10]), where robots are subject to *passive mobility*, also used in our paper. Passive mobility aims to model volatile environments like water flow, wind or unstable mobility of agents' carriers. Further, [10] considered different speed of such agents. (cf. also [16]).

A well-known methodology for distributed patrolling is the rotor router, which provides local (distributed) control at the nodes of an underlying graph for managing the movement of robots. The original study of patrolling based on rotor routers is due to [32] in which they also study idle and stabilization time of patrolling strategies. Additional work on rotor routers can be found in the work of [11,17] as well as [9,32], which consider the cover time of a single walk while [27] determines the precise asymptotic value of the rotor-router cover time. In our paper, the robots are autonomous and interact (changing direction) only when they meet, while convergence of the resulting distributed patrolling algorithms is established by analyzing a dynamical system.

1.2 Formal Model and Problem Definitions

Consider a set R of n mobile robots $r_1, \ldots, r_n$ each associated with some *patrolling speed* p_i and some *walking speed* w_i, where $p_i < w_j$ for $i, j = 1, \ldots, n$. Robots perpetually move along the unit segment $[0, 1]$ in both directions. At any moment, a robot r_i may be in a *walking state* in which it moves at its walking speed w_i or in the *patrolling state* and moving at its patrolling speed p_i. Each robot may walk in both directions but its patrolling is always done in the same direction of the interval $[0, 1]$. Hence we can say that each robot is associated with its patrolling direction (positive or negative direction of the interval) which does not change throughout its entire movement.

The Fence Patrolling Problem: Let S denote a *patrolling schedule*, i.e. an algorithm associated with each of the robots, instructing with what speed and in what direction every robot moves at each moment of time. Given a schedule produced by algorithm A, the idle time $I_A(P, t)$ of a point P at time t is defined as the amount of time needed to the next visit of P after time t by any robot of R:

$$I_A(P, t) = \min\{t_v > t : \exists i (0 \le i \le n \text{ such that } r_i(t_v) = P)\} - t,$$

where we denote by $r_j(t)$ the point of the segment visited by r_j at time t, for $t \in [0, \infty)$. We are interested in an algorithm minimizing the maximal idle time taken over all points P and all time moments t. However, as it may be impossible to design an algorithm which achieves the best possible idle time right from the start of the schedule (e.g. robots may be in an unsuitable position to achieve this) we will be allowing any finite delay for checking idle times of segment points. More exactly, the *idleness* of algorithm A is defined as $I_A = \inf_{T \ge 0} \sup_{0 \le p \le 1} I_A(P, T)$. An optimal patrolling algorithm is a schedule which optimizes the idleness among all possible patrolling algorithms under consideration.

Our goal may also be viewed as the following equivalent maximization task. Suppose that the robots operate in an infinite line. What is the maximal length segment $[0, L]$ and the perpetual movement of the robots of R, such that in each unit time interval $[t, t+1]$ each point of $[0, L]$ is visited at least once by a robot in patrolling state. We call such a length L the *patrolling range* of R.

The Distributed Model: In this model, robots have only very primitive capabilities allowing them to execute separate schedules which change according to their perception of the environment. They are *oblivious* and *silent* (cannot communicate) and they cannot process any information. Besides two-speed mobility they can perceive the environment by recognizing obstacles (i.e., endpoints) or other robots that they do not even need to recognize. They are not aware of their patrolling or walking speeds nor of the length of the segment. In the schedules produced by our distributed algorithms the robots function according to the *bouncing-rule*: if a robot collides with another robot or a segment endpoint, then it changes direction as well as moving-state. A patrolling schedule of n robots is in a *stable configuration* $(x_1^*, \ldots, x_n^*)$ if robot i moves within the interval $[x_{i-1}^*, x_i^*]$ (we set $x_0^* = 0$ and $x_n^* = 1$), bouncing always at its endpoints.

Scheduling with Regular Delivery: Our techniques allow us to solve the following transportation problem. Suppose that at point 1 of the unit segment there is an infinite quantity of a commodity that is transported to point 0 by robots R. Each robot may carry one item of the commodity using a speed not exceeding p_i or it may travel with no load with a speed not exceeding w_i. At any time, a robot r_i may drop the item it is carrying at the point currently occupied by r_i or it may pick up an item present at its current position. When two robots meet at a point an item being carried by one of them may be transferred to the other one. At all times no robot may carry more than one commodity item. What is the smallest value I, such that during any time interval $(kI, (k+1)I]$ a new item is delivered to point 0? Since the interval under consideration is of unit length, such a value of I would correspond to commodity flow of I units between the two endpoints.

1.3   Organization of the Paper and Summary of Contributions

We attempt to solve the Fence Patrolling Problem using centralized or distributed algorithms. As a warm-up, we give in Sect. 2 an optimal centralized algorithm for the problem, whose solution serves as a benchmark for subsequent sections. We conclude in Sect. 2.2 with an interesting application to an optimization transportation problem. Our main technical contributions appear in Sect. 3 where we study the Fence Patrolling Problem in a distributed setting and where robots have only primitive capabilities. First, in Sect. 3.1 we optimally solve Fence Patrolling with 2 primitive robots. In the same section, we also develop the main ideas we built upon for patrolling with an arbitrary number of robots. Then, in Sect. 3.2 we introduce a generic distributed solution for patrolling with primitive robots. Our solution induces a complex dynamical system, whose analysis is the main focus of the remaining sections. Section 3.3 proposes a technical and highly non-trivial analysis of the dynamics of primitive robots, and concludes with an efficient and analytic algorithm for deciding whether the system of robots converges to an optimal solution. Finally, Sect. 3.4 studies restricted, yet natural families of primitive robot collections that have the potential of inducing dynamic systems that converge to stable and optimal solutions. With non-trivial and technical arguments, we conclude by showing that special families of three and four primitive robots do solve the Fence Patrolling problem optimally.

2   Fence Patrolling Using Centralized Algorithms

In this section we give the general, centralized algorithm generating optimal patrolling schedules for the Fence Patrolling Problem for any number of robots. Subsequently, we show how to solve optimally the Scheduling with Regular Delivery problem.

2.1   The Optimal Centralized Algorithm

The high level idea of the optimal algorithm (Algorithm 1 below) is to carefully divide the unit interval into subsegments, and to make each robot operate only in the subsegment associated with it. In particular, robot r_i is associated with a subsegment of size proportional to $\frac{p_i w_i}{p_i + w_i}$. Each robot zigzags between the endpoints of its subsegment patrolling in one direction and walking in the opposite direction. As the following lemmata indicate, the subsegments are chosen so that the idle time for all points is the same, except from points patrolled by two robots. A formal description of algorithm follows.

First we calculate the idleness of the proposed algorithm. .

Lemma 1. *The idleness I_1 of Algorithm 1 satisfies $I_1 = 1/\sum_{i=1}^{n} \frac{1}{1/p_i + 1/w_i}$.*

Proof. As σ_j, $j = 0, \ldots, n$, is an increasing sequence, each robot r_j operates within the subsegment $[x_{j-1}, x_j]$, for $j = 1, \ldots, n$, having interior disjoint with all other subsegments. Consider any index j and any point $x \in [x_{j-1}, x_j]$. Denote by t^* a time when r_j is in the patrolling state and $r_j(t^*) = x$. Before r_j visits x in patrolling state the next time, it has to traverse from x to x_{j-1} patrolling followed by walking the subsegment $[x_{j-1}, x_j]$ and patrolling from x_j to x. The time needed for this equals

---

Algorithm 1. Centralized Schedule.

Input: n robots R with associated walking and patrolling speeds $w_i, p_i, i = 1 \ldots, n$, respectively.
Output: Schedule of R
1: Let $\sigma_j = \sum_{i=1}^{j} \frac{1}{1/p_i + 1/w_i}$ for $j = 1, \ldots, n$
2: **for** $j = 1, \ldots, n$ **do**
3:      Place robot r_j at initial position $x_j = \sigma_j / \sigma_n$
4:      **repeat forever**
5:          In patrolling state, move (left) at speed p_j until reaching point x_{j-1}
6:          In walking state, move (right) at speed w_j until reaching point x_j

---

$$\frac{x - x_{j-1}}{p_j} + \frac{x_j - x_{j-1}}{w_j} + \frac{x_j - x}{p_j} = (x_j - x_{j-1})\left(\frac{1}{p_j} + \frac{1}{w_j}\right) = \frac{1}{\sum_{i=1}^{n} \frac{1}{1/p_i + 1/w_i}}$$

Next we show that within certain time windows, robots' walked distance cannot be much less than their patrolled distance.l

Lemma 2. *Let $D > 1$ be the distance patrolled by robot r_i during some time interval. Then in the same time interval r_i must walk distance at least $D - 1$.*

Proof. Suppose, by symmetry, that r_i can patrol only in the right-to-left (i.e. negative) direction of the interval. The difference between the initial and the final position of the robot equals to $R - L$, where L denotes the sum of the lengths of its left-to-right moves and R denotes the sum of the lengths of its right-to-left moves. Observe that $R - L \leq 1$. As the total patrolling distance is at most R and the total walking distance is at least L we have the claim of the lemma.

We are now ready to prove that the Algorithm 1 is optimal.

Theorem 1. *Algorithm 1 produces a patrolling schedule with optimal idleness.*

Proof (of Theorem 1). Consider any algorithm $\mathcal{A}$ and its idleness $I_{\mathcal{A}}$. It is sufficient to show, that for any $\epsilon > 0$ there exists a time interval $T = [t^*, t^* + I_1 - \epsilon]$ during which some point of the segment is not patrolled by any robot. Let κ be an integer such that $\kappa > \dfrac{\sum_{i=1}^{n} \frac{1/w_i}{1/p_i + 1/w_i}}{\epsilon \sum_{i=1}^{n} \frac{1}{1/p_i + 1/w_i}}$. Consider any time interval K of size $= \kappa I_{\mathcal{A}}$. We prove that K must contain interval T with the property from the claim made above.

Let d_i denote the distance traversed by r_i while patrolling during interval K. As each point of the segment must be patrolled at least κ times during time interval K we have $\sum_{i=1}^{n} d_i \geq \kappa$. By Lemma 2 we have $\frac{d_i}{p_i} + \frac{d_i - 1}{w_i} \leq \kappa I_{\mathcal{A}}$, hence $d_i \leq \frac{\kappa I_{\mathcal{A}} + 1/w_i}{1/p_i + 1/w_i}$. Therefore

$$\kappa \leq \sum_{i=1}^{n} d_i \leq \kappa I_{\mathcal{A}} \sum_{i=1}^{n} \frac{1}{1/p_i + 1/w_i} + \sum_{i=1}^{n} \frac{1/w_i}{1/p_i + 1/w_i}$$

and

$$I_{\mathcal{A}} \geq \frac{1}{\sum_{i=1}^{n} \frac{1}{1/p_i + 1/w_i}} - \frac{\sum_{i=1}^{n} \frac{1/w_i}{1/p_i + 1/w_i}}{\kappa \sum_{i=1}^{n} \frac{1}{1/p_i + 1/w_i}} \geq \frac{1}{\sum_{i=1}^{n} \frac{1}{1/p_i + 1/w_i}} - \epsilon$$

This proves Theorem 1.

Note that due to Lemma 1, it follows that almost all points of the segment have the same idle time (with the exception of the endpoints of subsegments which may be alternately visited by two robots). From Algorithm 1, it follows that each robot r_i operates solely inside a subsegment of size $\frac{1}{1/p_i + 1/w_i} / \sum_{i=1}^{n} \frac{1}{1/p_i + 1/w_i}$, for $i = 1, \ldots, n$. By scaling down the consideration to the idle time of 1 we easily obtain:

Corollary 1. *The patrolling range of robot r_i having patrolling speed p_i and walking speed w_i equals $\frac{1}{1/p_i + 1/w_i}$. The patrolling range of a set of robots equals the sum of their patrolling ranges.*

2.2 Scheduling with Regular Delivery (an Application to Transportation)

In this section we show an interesting application of our previous findings in *Scheduling with Regular Delivery* (see Sect. 1.2 for the definition). Recall that, at a high level, the goal in this optimization problem is to maximize the flow of the commodity from one endpoint of the interval to the other. Intuitively, in an optimal solution, robots should be fully synchronized, so that in particular no robot stays idle at any moment. Interestingly, the patrolling schedule of Algorithm 1 has exactly this property. More specifically it is not difficult to see that since every point within each subsegment has the same idle time, robots are fully synchronized and return to their initial configuration/positioning with a period of I_1. Robots that move in patrolling mode should be understood as carriers of a commodity unit, while they are in walking mode if no commodity is carried. Also, when robots meet at the endpoints of a subsegment, they exchange the commodity that one of them is carrying. Note that within period I_1, every robot carries one unit of flow between the endpoints of its subsegment.

Proposition 1 below shows formally that the idleness of Algorithm 1 is the optimal value for the problem of Scheduling with Regular Delivery.

Proposition 1. *The solution (i.e., the optimal time interval) of the problem concerning Scheduling with Regular Delivery for the set of robots R equals I_1 - the idleness of Algorithm 1.*

Proof. The discussion above explains how a patrolling schedule corresponds to a feasible delivery schedule for Scheduling with Regular Delivery, in which the flow equals I_1. Next we argue that every delivery schedule with flow I also corresponds to a patrolling schedule with idle time I.

Indeed, consider some delivery schedule with flow I. That is, for every k, during any time interval $(kI, (k + 1)I)$ a new item is delivered to point 0. Consider an arbitrary point of the unit interval. Then it must be the case that at least one robot passes over that point at least once, carrying one unit of the commodity, between the time interval $(kI, (k + 1)I]$. Consider now a patrolling schedule in which robots patrol instead of carrying a commodity unit, and walk whenever no commodity units are carried (in the corresponding directions, respectively). The correspondence guarantees that the idle time of that point is no more than I.

3   Fence Patrolling with Distributed Robots

In this section we consider the case when the collection of patrolling robots acts in a distributed way (see Sect. 1.2 for the model). We are interested in an algorithm having the same idleness as the one of the optimal centralized algorithm even if our robots are very weak. For notational reference we will assume that there exist two motionless robots r_0 and r_{n+1} (i.e. $w_0 = p_0 = w_{n+1} = p_{n+1} = 0$), positioned at the left and right endpoint, respectively. This way every robot r_i, for $i = 1, \ldots, n$, bounces at its left or right neighbor. We have the following lemma.

Lemma 3. *For any collection R of robots there exists a centralized algorithm producing an optimal schedule in stable configuration, in which the robots behave according to the bouncing rule.*

Proof. Consider the partition of the unit interval as in line 1 of Algorithm 1. By Corollary 1 each robot executing Algorithm 1, within the same time interval, independently covers a segment subinterval proportional to its patrolling range. It is then possible to reschedule the robots' starting times so that each robot arrives at the left endpoint of its subinterval exactly at the same time as when its left neighbor arrives at the right endpoint of its interval, resulting in a meeting.

We will attempt to design a distributed algorithm producing robots trajectories converging to a schedule of a stable configuration of robots. The task seems of special interest, given that robots are assumed to be oblivious and silent.

3.1   Distributed Optimal Schedule for Two Primitive Robots

The purpose of this section is to demonstrate that two primitive robots can optimally solve the Fence Patrolling Problem.

Let I_{opt} be the optimal idleness of the offline schedule for two robots. We design an algorithm, for which for any $\epsilon > 0$ there exists a time t^*, such that in every time interval $[t, t + I_{opt} + \epsilon]$, with $t \geq t^*$, each point of the segment is visited by some robot. Obviously, using such weak robots, it is impossible to design an algorithm which achieves optimal idle time only after some finite time of their operation. This would need robots capable of recognizing the parameters of the environment (e.g. patrolling and walking speeds of robots, distance traveled, time between collisions, etc.). Since robots react only when colliding with each other, or when they reach one endpoint (as if they do not know when collisions will occur or where the endpoints are) we slightly abuse standard terminology and we call our algorithm online.

We show that Algorithm 2 is the optimal one, i.e. its idleness equals the idleness of the optimal offline Algorithm 1. The first critical observation is that collision points converge to a stable configuration.

Lemma 4. *The sequence of collision points of Algorithm 2 converges to the point*
$$\frac{\frac{1}{p_2} + \frac{1}{w_2}}{\frac{1}{p_1} + \frac{1}{w_1} + \frac{1}{p_2} + \frac{1}{w_2}}.$$

Algorithm 2. Online Schedule for Two Robots.

Input: Two robots r_1, r_2 placed at the two segment endpoints
Output: Schedule of R
 1: Both robots start in patrolling state moving towards each other.
 2: Each robot switches state and direction when colliding either with the other robot
    or with an endpoint.

Proof. Suppose that the two robots following the schedule produced by Algorithm 2 meet at a point x, $0 < x < 1$. Suppose also that before the meeting both robots were in the patrolling state and moving towards each other. We show first that the next meeting occurs at a point x' such that

$$x' = -\frac{\frac{1}{w_1} + \frac{1}{w_2}}{\frac{1}{p_1} + \frac{1}{p_2}} x + \frac{\frac{1}{p_2} + \frac{1}{w_2}}{\frac{1}{p_1} + \frac{1}{p_2}} \tag{1}$$

Indeed, as $p_1 < w_2$ and $p_2 < w_1$ no robot can be caught from behind by another one, while walking along the segment. Consequently, both robots reach the segment endpoints and they restart patrolling while moving towards each other, eventually colliding at x'. As both robots spend the same time while traveling from x to x' (cf. Fig. 1) we have $\frac{x}{w_1} + \frac{x'}{p_1} = \frac{1-x}{w_2} + \frac{1-x'}{p_2}$. Therefore $x' \left(\frac{1}{p_1} + \frac{1}{p_2} \right) = -x \left(\frac{1}{w_1} + \frac{1}{w_2} \right) + \frac{1}{p_2} + \frac{1}{w_2}$. We see from Identity (1) that $x' = -Ax + B$ where

$$A := \frac{\frac{1}{w_1} + \frac{1}{w_2}}{\frac{1}{p_1} + \frac{1}{p_2}}, \quad B := \frac{\frac{1}{p_2} + \frac{1}{w_2}}{\frac{1}{p_1} + \frac{1}{p_2}}. \tag{2}$$

If x_1 is the initial collision point then we see that the kth collision point satisfies the recurrence $x_k = -Ax_{k-1} + B$. This yields a geometric series from which it follows that $x_k = (-A)^k x_0 + B\frac{1-(-A)^k}{1+A}$. Since $A < 1$ as $k \to \infty$ we get $x_k \to \frac{B}{1+A} = \frac{\frac{1}{p_2} + \frac{1}{w_2}}{\frac{1}{p_1} + \frac{1}{w_1} + \frac{1}{p_2} + \frac{1}{w_2}}$ which completes the proof.

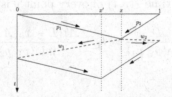

Fig. 1. Two robots r_1, r_2 start by patrolling in opposite directions. They first collide at a point x and they both change to walking. After bouncing (not necessarily at the same time) at the endpoints of the interval $[0, 1]$ they change to patrolling and meet anew at point x'. The vertical line indicates time.

Our intention is to generalize the lemma above for dynamic systems with arbitrarily many robots. Till then, we show that Algorithm 2 produces the optimal schedule.

Theorem 2. *The idleness of the schedule of Algorithm 2 equals I_1 - the idleness of the optimal schedule produced by the centralized algorithm.*

Proof. It is sufficient to show that for any $\epsilon > 0$ there exists a time moment t^* such that for any $t > t^*$ and any time interval $T = [t, t + I_1 + \epsilon]$ each point of the segment is patrolled by some robot. Observe that from the start to the first meeting point x_0 robot r_1 patrols the interval $[0, x_0]$, while during the same time interval r_2 patrols the interval $[x_0, 1]$. Hence we have $\frac{x_0}{p_1} = \frac{1-x_0}{p_2}$. Solving for x_0 we get $x_0 = \frac{1}{p_2} / \left(\frac{1}{p_1} + \frac{1}{p_2} \right)$. Observe that the distance $D = |x_k - x_{k-1}|$ between the $(k-1)$-th and k-th meeting points (defined in the proof of Lemma 4) we have

$$D = \left| \left((-A)^k x_0 + B\frac{1-(-A)^k}{1+A} \right) - \left((-A)^{k-1} x_0 + B\frac{1-(-A)^{k-1}}{1+A} \right) \right|$$

$$= \left| ((-A)^k - (-A)^{k-1}) \left(x_0 - \frac{B}{1+A} \right) \right|$$

$$= |(-A)^k (B - x_0(A+1))|$$

$$= A^k (B - x_0(A+1)) \qquad (\text{as } A > 0 \ \& \ (B - x_0(A+1)) > 0)$$

Since $|A| < 1$, D is converging to 0. As x_k and x_{k-1} are on the different sides of the convergence point $\frac{B}{1+A}$, within every time interval $I_1 + \frac{D}{\min(p_1,p_2)}$ the entire segment is jointly patrolled by both robots. Let K be such that $K \geq \log_A \frac{\epsilon \cdot \min(p_1,p_2)}{B-x_0(A+1)}$ and let

$$t^* > (K+1)(\frac{1}{p_1} + \frac{1}{w_1} + \frac{1}{p_2} + \frac{1}{w_2}).$$

As $(\frac{1}{p_1} + \frac{1}{w_1} + \frac{1}{p_2} + \frac{1}{w_2})$ is the time between two consecutive bounces between robots r_1, r_2, after time t^* the robots bounced at least k times. Hence (as $0 < A < 1$) for $t > t^*$ the idle time $I(p, t)$ of every point p is

$$I(p,t) \leq I_1 + \frac{D}{\min(p_1, p_2)} = I_1 + \frac{A^k(B - x_0(A+1))}{\min(p_1, p_2)} \leq I_1 + \epsilon,$$

which proves the theorem.

Observe that Algorithm 2 works even if the robots do not necessarily start their respective schedules at the same time. Indeed, because $p_i < w_j$, $i, j = 1, 2$, when the second robot wakes up and starts patrolling it cannot meet the robot which started first while this robot is in the walking state. Therefore the robots meet when they are both in the patrolling state and the subsequent meetings converge to the same point as before.

A Suboptimal Schedule

Note that, when robots start walking rather than patrolling, their subsequent meeting points do not converge. This follows from the proof of Theorem 2, as in this case, the critical value of quantity $A = \frac{1/p_1 + 1/p_2}{1/w_1 + 1/w_2} > 1$ (the roles of patrolling and walking speeds are swapped in the equation for x'). On the other hand if one robot starts in the walking state while the other one in the patrolling state the convergence is possible in at most one of the two symmetric cases, depending which among the two values $1/p_1 + 1/w_2$ or $1/w_1 + 1/p_2$ is larger. In particular, when $1/p_1 + 1/w_2 = 1/w_1 + 1/p_2$ (e.g. in the case of identical robots) the meeting points alternate between two symmetric positions on the segment and the idleness is clearly suboptimal (cf. Fig. 2).

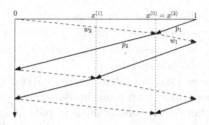

Fig. 2. Two robots r_1, r_2 perpetually alternate between $x(0)$ and $x(1)$ (r_1 starts by patrolling and r_2 by walking). Here $w_1 = 5, p_1 = 3\frac{1}{7}, w_2 = 10, p_2 = 4$.

3.2 Distributed Schedule for n Primitive Robots

In this section we propose a distributed solution for an arbitrary number of primitive robots that induces an interesting dynamic system, which we analyze in subsequence sections.

As observed for the case of two robots, the convergence was attested if the two robots were colliding always in the patrolling state. For more than two robots this condition can no longer be guaranteed as all but two robots collide with both neighbors. We begin our analysis by proposing Algorithm 3, an intuitive distributed schedule that assumes the bouncing rule, i.e. that robots can respond to bounces by flipping their moving state (e.g. from walking to patrolling) and their moving direction. Moreover, as in our setting robots will start walking simultaneously at the same segment endpoint we naturally assume that $p_i \neq p_j$ (and that $w_i \neq w_j$), otherwise identical robots would always stay together and only one of them would contribute to the patrolling algorithm.

Algorithm 3. Distributed Schedule for Many Robots.

Input: A collection R of robots with distinct patrolling and walking speeds
Output: Schedule of R
  1: All robots start from the rightmost endpoint of the interval, in patrolling state moving right-to-left.
  2: Each robot switches state and direction while colliding with the other robot or with an endpoint.

As all robots patrol in the same direction and they change states only when meeting we can conclude with the following:

Observation 3. Algorithm 3 produces a dynamic schedule with Regular Delivery, for a given set of robots.

Notice that our algorithm defines a complex dynamical system of memoryless robots moving back and forth in an interval. The analysis of the system dynamics is very complicated, given that robot collisions might occur either between robots moving in opposite directions, or in the same directions (i.e. a collision may happen from behind). In what follows we analyze the dynamics under the assumption that we have one type of collisions, which as we shall see in the next subsections naturally arise by restricting the configuration of the robot speeds to what we later call monotone speeds.

We call the dynamical system that arises from Algorithm 3 *regular* if collisions occur only between robots that move in opposing directions, and therefore collisions happen only between a robot that is in patrolling state moving right-to-left and a robot in walking state moving left-to-right. We are interested in answering whether regular dynamical systems converge to a stable configuration, and whether this configuration has optimal idle time. Below we propose a highly efficient algorithm for verifying whether a regular dynamical system has this property (for any number of robots). Then we answer this question in the positive for up to 4 robots, under the assumption that speeds satisfy a natural condition.

3.3 Regular Systems Dynamics of Primitive Robots

Dynamic systems of primitive robots induced by Distributed Algorithm 3 are highly complex and difficult to analyze. The purpose of this section is to provide a deep and technical analysis of the dynamics of regular systems. We conclude the section by proposing a highly non-trivial algorithm for deciding convergence of primitive robots in a stable configuration (where robots eventually move within disjoint subintervals).

Notice that in every dynamical system, and at every point in the time horizon, robots will appear on the interval in the same order. We rename the robots so that robot $i + 1$ is always to the right of robot i, after they develop according to Algorithm 3. Below we denote by x_t^i the point in the interval where robots r_i, r_{i+1} bounce for the t-th time. The purpose of the next lemma is to predict points x_t^i. The reader may view it as the analogue of (part of the proof of) Lemma 4 that dealt with only two robots.

Lemma 5. *In a regular system we have*

$$\left(\frac{1}{p_{i+1}} + \frac{1}{w_i}\right) x_{t+1}^i - \left(\frac{1}{p_i} + \frac{1}{w_i}\right) x_{t+1}^{i-1} = \left(\frac{1}{p_{i+1}} + \frac{1}{w_{i+1}}\right) x_t^{i+1} - \left(\frac{1}{p_i} + \frac{1}{w_{i+1}}\right) x_t^i$$

Proof. First we claim that $x_t^{i-1} < x_t^i$, $i = 1, \ldots, n$, and that robots r_{i-1}, r_i bounce at x_t^{i-1} before r_i, r_{i+1} bounce at x_t^i. Indeed, let $\tau(x_t^i)$ denote the time of this bounce. Note that robot r_1 (while patrolling) bounces first at the origin and on its way back (now walking) bounces with robot r_2 (which is patrolling). After robot r_2 bounces with robot r_1, it begins walking and eventually bounces with robot r_3 which moves in opposite direction and is patrolling, etc. This reasoning shows that $x_1^{i-1} < x_1^i$, $i = 1, \ldots, n$ as well as that r_{i-1}, r_i bounce for the first time before r_i, r_{i+1} do. Since the system is regular, each robot alternates its bounces between both neighbors r_{i-1}, r_{i+1} which is sufficient to conclude the claim.

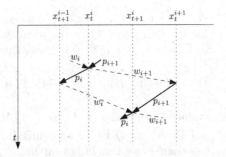

Fig. 3. Two time consecutive bounces between robots r_i, r_{i+1}.

Next observe that between time $\tau(x_t^i)$ and $\tau(x_{t+1}^i)$ robot r_{i-1} first patrols right-to-left the interval $[x_{t+1}^{i-1}, x_t^i]$ then it walks left-to-right the interval $[x_{t+1}^{i-1}, x_{t+1}^i]$. During the same time interval r_i first walks left-to-right the interval $[x_t^i, x_t^{i+1}]$ then it patrols right-to-left the interval $[x_{t+1}^i, x_t^{i+1}]$ (see Fig. 3). Comparing both times we get $\frac{x_t^i - x_{t+1}^{i-1}}{p_i} + \frac{x_{t+1}^i - x_{t+1}^{i-1}}{w_i} = \frac{x_t^{i+1} - x_t^i}{w_{i+1}} + \frac{x_t^{i+1} - x_{t+1}^i}{p_{i+1}}$. Regrouping terms implies the lemma.

We can rewrite the recurrence in Lemma 5 in a more concise matrix form. Define the following $(n-1) \times (n-1)$ matrices A, B and vector $c \in \mathbb{R}^{n-1}$:

$$A = \begin{pmatrix} {}^1\!/_{p_2} + {}^1\!/_{w_1} & 0 & 0 & \cdots & 0 & 0 \\ -{}^1\!/_{p_2} - {}^1\!/_{w_2} & {}^1\!/_{p_3} + {}^1\!/_{w_2} & 0 & \cdots & 0 & 0 \\ 0 & -{}^1\!/_{p_3} - {}^1\!/_{w_3} & {}^1\!/_{p_4} + {}^1\!/_{w_3} & \cdots & 0 & 0 \\ \vdots & \vdots & \vdots & \ddots & \vdots & \vdots \\ 0 & 0 & 0 & \cdots & {}^1\!/_{p_{n-1}} + {}^1\!/_{w_{n-2}} & 0 \\ 0 & 0 & 0 & \cdots & -{}^1\!/_{p_{n-1}} - {}^1\!/_{w_{n-1}} & {}^1\!/_{p_n} + {}^1\!/_{w_{n-1}} \end{pmatrix}$$

$$B = \begin{pmatrix} {}^1\!/_{p_1} + {}^1\!/_{w_2} & -{}^1\!/_{p_2} - {}^1\!/_{w_2} & 0 & \cdots & 0 & 0 \\ 0 & {}^1\!/_{p_2} + {}^1\!/_{w_1} & -{}^1\!/_{p_3} - {}^1\!/_{w_3} & \cdots & 0 & 0 \\ 0 & 0 & {}^1\!/_{p_3} + {}^1\!/_{w_4} & \cdots & 0 & 0 \\ \vdots & \vdots & \vdots & \ddots & \vdots & \vdots \\ 0 & 0 & 0 & \cdots & {}^1\!/_{p_{n-2}} + {}^1\!/_{w_{n-1}} & -{}^1\!/_{p_{n-1}} - {}^1\!/_{w_{n-1}} \\ 0 & 0 & 0 & \cdots & 0 & {}^1\!/_{p_{n-1}} + {}^1\!/_{w_n} \end{pmatrix}$$

$$c^T = \begin{pmatrix} 0 & 0 & \cdots & 0 & {}^1\!/_{p_n} + {}^1\!/_{w_n} \end{pmatrix}.$$

Theorem 4. *Consider a regular dynamical system of n robots (produced by Algorithm 3) and let A, B, c be the matrices defined in equations above. If the moduli (norms) of all eigenvalues of the matrix $A^{-1}B$ are less than 1, then the schedule of Algorithm 3 converges to a schedule in stable configuration which is also optimal (w.r.t. to centralized algorithms). In particular, for every $\epsilon > 0$, after $\Theta(\log 1/\epsilon)$ bounces (iterations) of any pair of neighboring robots the idle time $I_3(p, t)$ is such that $I_3(p, t) \leq (1 + \epsilon) \frac{1}{\sum_{i=1}^n \frac{1}{1/p_i + 1/w_i}}$.*

Proof. Let $X_t \in \mathbb{R}^{n-1}$ be the vector $\left(x_t^1, \ldots, x_t^{n-1}\right)^T$, where x_t^i is the bouncing point of robots r_i, r_{i+1} for the t-th time. Then the recurrence of Lemma 5 can be rewritten in matrix form as $AX_{t+1} + BX_t = c$. From this, we derive that

$$X_t = (-1)^t \left(A^{-1}B\right)^t + \left(I + A^{-1}B\right)^{-1} \left(I - (-1)^{-1}\left(A^{-1}B\right)^t\right) A^{-1}c.$$

Next consider the eigenvalue decomposition $A^{-1}B = Q\Lambda Q^T$, where Q is an orthogonal matrix. Then $\left(A^{-1}B\right)^t = Q\Lambda^t Q^T$, and since $\lim_{t\to\infty} \Lambda^t = \mathbf{0}$ as all eigenvalues have norm less than 1, we conclude that $\lim_{t\to\infty} X_t$ exists, i.e. the sequence converges to $X^* = \left(I + A^{-1}B\right)^{-1} A^{-1}c = (A+B)^{-1}c$ and the convergence is linear. From the definition of the recurrence, it follows that the schedule of Algorithm 3 converges to the schedule S which is in stable configuration X^*.

By Corollary 1 which, in view of Lemma 3, applies also to stable configurations the patrolling range of the collection of robots equals $\sum_{i=1}^{n} \frac{1}{1/p_i+1/w_i}$ and by the rate of convergence, after $\Theta\left(\log 1/\epsilon\right)$ bounces of neighboring robots, the idle time is already no more than than $(1+\epsilon)\frac{1}{\sum_{i=1}^{n} \frac{1}{1/p_i+1/w_i}}$.

Note that already Theorem 4 suggests a numerical method for checking whether the dynamical system arising from Algorithm 3 converges or not; given patrolling and walking speeds p_i, w_i, first compute matrix A^{-1}, and then calculate all eigenvalues of $A^{-1}B$ and verify that their norm is less than 1. Matrix inversing can be done explicitly and efficiently, say by Gauss-Jordan elimination or by LU decomposition. Finding however the eigenvalues of the non-Hermitian matrix $A^{-1}B$ is at least as difficult as finding the roots of high degree polynomials. In light of Abel's impossibility theorem, one has to rely on numerical methods to verify that the moduli of the eigenvalues of $A^{-1}B$ are indeed less than 1. In fact, a number of sophisticated numerical methods have been proposed to efficiently find eigenvalues of special families of matrices.

We depart from this approach, and in contrast to numerical methods, we propose an explicit, symbolic and efficient algorithm for verifying the precondition of Theorem 4 without explicitly computing the eigenvalues of $A^{-1}B$. Our strategy is to first give an explicit expression of A^{-1}, which in turn will allow us to calculate the characteristic polynomial of $A^{-1}B$. Finally, we invoke a powerful theorem that characterizes the range of polynomial roots (without finding them), and that can be exploited algorithmically.

We begin by calculating the characteristic polynomial of $A^{-1}B$, which will be useful also in subsequent subsections. Note that for a group R of n robots, the characteristic polynomial of $A^{-1}B$ is of degree $n - 1$. Also, any $r \times r$ leading principal minor $A^{-1}B$ can be computed from the $r \times r$ leading principal minors of A, B, which only depend on robots $1, \ldots, r + 1$. In fact the $r \times r$ leading principal minor $A^{-1}B$ is exactly the critical matrix whose eigenvalues determine the convergence of Algorithm 3 for input robots $1, \ldots, r+1$. So, we are motivated in denoting by $D_r(\lambda)$ the characteristic polynomial of the $r \times r$ leading principal minor $A^{-1}B$ (i.e. $D_{n-1}(\lambda) = |A^{-1}B - \lambda I|$). We choose to abbreviate $D_r(\lambda)$ by D_r. The next lemma provides two alternative recursive relations for D_r (each will be convenient in different arguments) that allow us to compute D_{n-1}, and will be also used later to establish convergence for special cases of robots. For notational convenience, we introduce the following abbreviations for some expressions that involve speeds w_i, p_i, and that appear in the definition of matrices A, B:

$$a_i := \frac{1}{p_{i+1}} + \frac{1}{w_i}, \quad b_i := \frac{1}{w_{i+1}} + \frac{1}{p_i}, \quad d_i := -\frac{1}{p_{i+1}} - \frac{1}{w_{i+1}},$$

With that notation, matrices A, B are written as

$$A = \text{diag}\,(a_i)_{i=1,\ldots,n-1} - \text{L-diag}\,(d_{i-1})_{i=2,\ldots,n-1}$$
$$B = \text{diag}\,(b_i)_{i=1,\ldots,n-1} - \text{U-diag}\,(d_{i-1})_{i=2,\ldots,n-1}$$

where by L-diag and U-diag we denote the low and upper diagonal matrices of dimension $(n-1) \times (n-1)$ and entries as indicated placed below and above the main diagonal, respectively.

Lemma 6. *For the characteristic polynomials D_r, the following equivalent recurrences hold for all $r \geq 2$*

$$D_r = (\alpha_r - \lambda) \cdot D_{r-1} + \sum_{t=1}^{r-2} \left(\alpha_{t+1} \cdot \prod_{j=t+2}^{r} \beta_j \right) \cdot D_t + \frac{b_1}{a_1} \prod_{j=2}^{r} \beta_j, \quad (3)$$

$$D_r = \left(\frac{b_r}{a_r} - \lambda \right) D_{r-1} + \frac{c_{r-1}^2}{a_{r-1} a_r} \lambda D_{r-2} \quad (4)$$

where $\alpha_i := \frac{b_i}{a_i} - \frac{d_{i-1}^2}{a_{i-1} a_i}, \beta_i := \frac{d_{i-1}^2}{a_{i-1} a_i}$. and with initial conditions $D_1 = \frac{b_1}{a_1} - \lambda, \ D_0 = 1$.

Proof. Define the $(n-1) \times (n-1)$ matrix K whose elements are

$$A_{ij}^{-1} := (-1)^{i+j} \frac{\prod_{t=j}^{i-1} d_t}{\prod_{r=j}^{i} a_r}, \quad \text{if } j \leq i$$

and is 0 otherwise. We claim that $A^{-1} = K$.

Indeed, note that by definition of A, we have that $(AK)_{i,j} = \sum_{k=1}^{n-1} A_{i,k} K_{k,j} = A_{i,i-1} K_{i-1,j} + A_{i,i} K_{i,j}$. So, when $j < i$, we have

$$(AK)_{i,j} = d_{i-1}(-1)^{i+j-1} \frac{\prod_{t=j}^{i-2} d_t}{\prod_{r=j}^{i-1} a_r} + a_i(-1)^{i+j} \frac{\prod_{t=j}^{i-1} d_t}{\prod_{r=j}^{i} a_r}$$

$$= (-1)^{i+j-1} \frac{\prod_{t=j}^{i-2} d_t}{\prod_{r=j}^{i-1} a_r} \left(d_{i-1} - a_i \frac{d_{i-1}}{a_i} \right) = 0.$$

Second, it is much easier to see that $(AK)_{i,j} = 0$ if $j > i$. And finally, we have

$$(AK)_{i,i} = a_i(-1)^{2i} \frac{1}{a_i} = 1,$$

concluding that $A^{-1} = K$ as promised (note also that K is a lower triangular matrix).

Therefore, for the so-called Hessenberg matrix $A^{-1}B$ (as it has zero entries above the first subdiagonal) we have that

$$\left(A^{-1}B\right)_{ij} = \sum_{k=1}^{n-1} K_{i,k}B_{k,j} = (-1)^{i+j}\left(\frac{b_j}{a_i}\prod_{t=j}^{i-1}\frac{d_t}{a_t} - \frac{d_{j-1}}{a_i}\prod_{t=j}^{i}\frac{d_{t-1}}{a_{t-1}}\right)$$

if $i \geq j - 1$, and 0 otherwise, with the understanding that $d_0 = 0$.

The following interesting relation holds for the entries of $A^{-1}B$, that is useful in finding the characteristic polynomial of the matrix.

$$\left(A^{-1}B\right)_{i,j} = -\frac{d_{i-1}}{a_i}\left(A^{-1}B\right)_{i-1,j}, \quad \forall i > j. \tag{5}$$

Next we introduce D'_r to denote a small variation of D_r. D'_r is the determinant of the same principal minor of $A^{-1}B - \lambda I$ (up to entry (r, r)) with the only difference that the entry (r, r) is replaced by $\left(A^{-1}B\right)_{r,r}$, instead of $\left(A^{-1}B - \lambda I\right)_{r,r}$.

With this notation, we can evaluate $\left|A^{-1}B - \lambda I\right|$ by expanding the determinant with respect to the entries $(n - 1, n - 1)$ and $(n - 2, n - 1)$. Using Eq. (5), we observe that

$$D_r = \left(\frac{b_r}{a_r} - \frac{d_{r-1}^2}{a_{r-1}a_r} - \lambda\right)D_{r-1} + \frac{d_{r-1}^2}{a_{r-1}a_r}D'_{r-1}, \tag{6}$$

$$D'_r = \left(\frac{b_r}{a_r} - \frac{d_{r-1}^2}{a_{r-1}a_r}\right)D_{r-1} + \frac{d_{r-1}^2}{a_{r-1}a_r}D'_{r-1}, \tag{7}$$

where the recurrence ends at $D_1 = \frac{b_1}{a_1} - \lambda$ and $D'_1 = \frac{b_1}{a_1}$. Repeated substitution of (7) to (6) and some direct calculations imply recurrence (3). Recurrence (4) is obtained from (3) by subtracting two consecutive terms of the sequence D_r.

Notice that Lemma 6, and in particular Eq. (4), allows us to calculate the characteristic polynomial D_{n-1} of $A^{-1}B$ by performing no more than $\Theta(n^2)$ arithmetic operations (additions, multiplications and divisions) between speeds p_i, w_i. Next we give an efficient algorithm for deciding whether the moduli of the roots of an arbitrary polynomial $f : \mathbb{R} \mapsto \mathbb{R}$ are all less than 1. Our intention is to run Algorithm 4 with input D_{n-1}, i.e. the characteristic polynomial of $A^{-1}B$.

Theorem 5. *A set of n robots R for which the output of Algorithm 3 gives a regular dynamical system converges to a stable configuration if and only if Algorithm 4 outputs YES on input D_{n-1}. As a result, convergence can be decided in $\Theta(n^2)$ arithmetic operations.*

Proof. By Theorem 4, the regular dynamical system converges to a stable configuration if and only if all eigenvalues of $A^{-1}B$ have moduli less than 1. The characteristic polynomial of $A^{-1}B$ can be computed in $\Theta(n^2)$ many operations,

---

Algorithm 4. Decide Convergence.

Input: A polynomial $f : \mathbb{R} \mapsto \mathbb{R}$ of degree t of the form $\sum_{i=0}^{t} \gamma_i \lambda^i$

1: Set $\gamma_i^{(0)} = \gamma_i$, for $i = 0, \ldots, t$.

2: For $j = 0, \ldots, t-1$ and for $k = 0, \ldots, j+1$ compute $\gamma_k^{(j+1)} = \gamma_0^{(j)} \gamma_k^{(j)} - \gamma_{n-j}^{(j)} \gamma_{n-j-k}^{(j)}$.

3: Compute $\delta_{j+1} := \gamma_0^{(j+1)} = \left(\gamma_0^{(j)} \right)^2 - \left(\gamma_{n-j}^{(j)} \right)^2$ for $j = 0, \ldots, t-1$.

Output: YES if and only if $\delta_1 < 0$ and $\delta_j > 0$ for $j = 2, \ldots, t$.

---

as a corollary of Lemma 6. Clearly, Algorithm 4 requires no more than $\Theta(n^2)$ arithmetic operations. Therefore, we can decide convergence in $\Theta(n^2)$ arithmetic operations as long as we can show that Algorithm 4 correctly decides whether the input polynomial f has all its roots (real or complex) strictly inside the unit circle.

Correctness of Algorithm 4 is an immediate corollary of Theorem 42,1, p. 150 in [29]: "Set $\Delta_r = \prod_{j=1}^{t} \delta_j$, , $r = 1, \ldots, t$ and suppose that k many of the products Δ_r are negative, and the remaining $t - k$ of them are positive. Then f has exactly k roots strictly inside the unit circle, exactly $t - k$ roots strictly outside the unit circle (and hence no roots on the unit circle)."

3.4 Monotone Robot Collections, and Convergence

In this section we demonstrate some special families of primitive robots that can solve Fence Patrolling optimally. The analysis even of the restricted families of three or four robots remains surprisingly technical and non-trivial.

Our technical results of Sect. 3.3 on regular dynamical systems raise the question whether such systems exist. A natural family of robots is when either the sum or the product of patrolling and walking speeds is constant for all robots or when some constant "power" of a robot may be used for improving its patrolling ability at the expense of its walking ability. In such a collection of robots, all patrolling speeds are dominated by the walking speeds, and the non-increasing order of patrolling speeds is the inverse order of that of the walking speeds. We make the definition formal.

Definition 1. *The collection R of n robots is called monotone if for $i, j = 1, \ldots, n$: 1) $p_i < w_i$, 2) $p_i \neq p_j$, and 3) $p_i < p_j \implies w_i > w_j$.*

A natural example of a monotone collection of robots is one where each robot i independently decides how to distribute it's energy e, which is the same for all robots, to walking and patrolling speeds w_i, p_i respectively, such that $w_i + p_i = e$. As it is observed before, a collection of robots that develop according to Algorithm 3 preserve the order they appear on the line. Without loss of generality we may assume that their indices are consecutive along the segment, i.e. that $w_n > w_{n-1} > \cdots > w_1 > p_1 > p_2 > \cdots > p_n$.

Lemma 7. *For a monotone collection of robots R, the dynamical system that arises from Algorithm 3 is regular (i.e. collisions occur while robots approach each other, the left one being in the walking state and the right one in the patrolling state.)*

Proof. Initially all robots walk right-to-left until the fastest walking robot collides with the left endpoint and starts walking left-to-right. Any "head on" collision results in the right robot switching to patrolling left-to-right and the left robot switching to patrolling right-to-left. So it is sufficient to prove that collisions from behind never take place. Suppose to the contrary, that there exists such a collision between a pair of consecutive robots on the segment, r_i, r_{i+1}, for $i = 1, \ldots, n-1$. Obviously r_i, r_{i+1} cannot collide when r_i moves left and r_{i+1} moves right. If both robots move right-to-left then, by assumption, they must be walking, and since $w_i < w_{i+1}$, r_i cannot catch r_{i+1}. Similarly, if both robots move left-to-right, by assumption, they are patrolling and since $p_i > p_{i+1}$, r_{i+1} cannot catch r_i.

As an immediate observation, we also obtain that

$$\frac{b_i}{a_i} < 1 \quad \& \quad \frac{(d_{i-1})^2}{a_{i-1}a_i} < 1 \tag{8}$$

for all $i = 1, \ldots, n-1$ and $i = 2, \ldots, n-1$ respectively, and for all regular collections of n robots, where a_i, b_i, c_i are as in Lemma 6. In fact, the characteristic polynomial D_{n-1} has leading coefficient $(-1)^n$, while it is also immediate from (4) that the constant coefficient is $\prod_{i=1}^{n-1} \frac{b_i}{a_i} < 1$. This automatically shows that the condition $\delta_1 < 0$, of Algorithm 4, holds true for monotone collections of robots. In fact, we conjecture that monotone collections of robots always converge to a stable configuration, i.e. that $\delta_j > 0$ for $j = 2, \ldots, n-1$, but a general proof is eluding us. Still the proof of convergence for up to $n \le 3$ robots is possible to establish. The proof of the next proposition relies on (8).

Proposition 2. *For a monotone collection of $n \le 3$ robots the schedule produced by Algorithm 3 has the optimal idleness $I_3 = 1/\sum_{i=1}^{n} \frac{1}{1/p_i + 1/w_i}$.*

Proof. We show that the conditions of Theorem 4 are satisfied. The case of 1 robot is straightforward. For $n = 2$ robots, the characteristic polynomial is $D_1 = \frac{b_1}{a_1} - x$, which by (8) has one real root with absolute value less than 1.

Now we turn our attention to $n = 3$, and by (4) the characteristic polynomial $D_2(\lambda)$ has the form $P_W(\lambda) := (U - \lambda)(V - \lambda) + W\lambda = \lambda^2 - (U + V - W)\lambda + UV$, where U, V, W are non negative constants which by (8) are strictly less than 1.

We show that the moduli of the roots of $P_W(\lambda)$ are less than 1. First we observe that $P_0(\lambda)$ has this property.

Case 1: If $P_W(\lambda)$ has real roots, then these are $\rho_{1,2}^{(W)} = \frac{U+V-W \pm \sqrt{(U+V-W)^2 - 4UV}}{2}$, with the understanding that $\rho_1^{(W)}, \rho_2^{(W)}$ correspond to the square root having positive and negative sign respectively. Then we observe that $\rho_1^{(W)} < \rho_1^{(0)} < 1$, while also $\rho_1^{(W)} > -W/2 > -1/2$ (by ignoring the positive terms). Hence $-1/2 < \rho_1^{(W)} < 1$. Similarly, we see that $\rho_2^{(W)} < (U + V)/2 < 1$ (by ignoring the negative terms). And finally note that $U + V - \sqrt{(U + V - W)^2 - 4UV} > -W$, since $(U + V + W)^2 > (U + V - W)^2 - 4UV$. Therefore, $\rho_2^{(W)} > (-2W)/2 = -1$, concluding that $-1 < \rho_2^{(W)} < 1$, as well.

Case 2: If $P_W(\lambda)$ has complex roots, say σ_1, σ_2, then it must be the case that $\|\sigma_1\|^2 = \|\sigma_2\|^2 = \sigma_1\sigma_2 = UV < 1$.

We now prove convergence of monotone collections of $n = 4$ robots by using a refinement of monotonicity.

Definition 2. *The collection R of n robots is called strongly monotone if it is monotone and for all r we have* $\left(\frac{1}{p_r} + \frac{1}{w_{r+1}}\right) \left(\frac{1}{p_r} + \frac{1}{w_{r-1}}\right) > \left(\frac{1}{p_r} + \frac{1}{w_r}\right)^2$.

Examples of strictly monotone robots are also common. As before, consider a collection of robots for which the product of their speeds is invariant, say that $p_r w_w = e$. Also, suppose that $w_r = rt$ for some constant $t > \sqrt{e}$, i.e. suppose that walking speeds form an arithmetic progression. For such a collection of robots we see that

$$\left(\frac{1}{p_r} + \frac{1}{w_{r+1}}\right) \left(\frac{1}{p_r} + \frac{1}{w_{r-1}}\right) - \left(\frac{1}{p_r} + \frac{1}{w_r}\right)^2 = \frac{e + 2r^2 t^2}{er^2 \left(r^2 - 1\right) t^2} > 0$$

as promised. Also, it is easy to see that $w_1 > p_1$.

Due to the definitions of a_i, b_i and that of α_i in Lemma 6, asking that a collection of robots is strongly monotone is equivalent to asking that $\alpha_r > 0$ for every r (see (3)). This allows us to show that characteristic polynomials associated with such robots have no negative real roots. We can now prove that the characteristic polynomial of every strongly monotone collection of robots has real roots less than 1 in absolute value.

Theorem 6. *For every monotone (not necessarily strongly) collection of n robots, $D_r(\lambda)$ preserves sign for all $\lambda \geq 1$. If in addition robots are strongly monotone, then $D_r(\lambda)$ preserves sign (and is actually positive) for all $\lambda < 0$. As a result all real roots of D_r lie strictly between 0 and 1.*

Proof. The less technical proof concerns the strongly monotone collections of robots. For this consider the cone $\mathcal{C}$ of polynomials of the form $\sum_{t=0}^{r}(-1)^t \rho_t \lambda^t$, where $\rho_t > 0$, i.e. polynomials whose odd-degree monomial coefficients are negative, and whose even-degree monomial coefficient are positive. Clearly, any polynomial $p(\lambda) \in \mathcal{C}$ is positive for every $\lambda < 0$ (and is actually decreasing).

We claim that for all $r \geq 0$, $D_r \in \mathcal{C}$. To that end, we first observe that the statement is true for $r = 0, 1$. For any $r \geq 2$, we invoke (3). Since all α_i, β_j are positive reals, we can show that $D_r \in \mathcal{C}$ as long as we can verify that $(\alpha_r - \lambda) \cdot D_{r-1} \in \mathcal{C}$ (the rest of summands in (3) are conical combinations of polynomials in $\mathcal{C}$). It is straightforward now to check that $-\lambda D_{r-1} \in \mathcal{C}$, hence $(\alpha_r - \lambda) \cdot D_{r-1} = \alpha_r D_{r-1} + (-\lambda D_{r-1}) \in \mathcal{C}$, as wanted.

Now we focus on a monotone (not necessarily strongly) collection of robots. We prove by induction on r that for all $\lambda \geq 1$, D_r is a polynomial which is

$$\begin{cases} \text{positive and increasing, if } r \text{ is even} \\ \text{negative and decreasing, if } r \text{ is odd} \end{cases}$$

Indeed, the statement is true for $r = 1, 2$. Next we turn our attention to $r \geq 3$. We have in mind to invoke (4). Now fix any $\lambda_0 \geq 1$. Note that $\frac{b_r}{a_r} - \lambda_0 < 0$. Next observe that if r is even, then $D_{r-1}(\lambda_0) < 0$ and $D_{r-2}(\lambda_0) > 0$, so that $D_r(\lambda_0) > 0$. Similarly, if r is odd, then $D_{r-1}(\lambda_0) > 0$ and $D_{r-2}(\lambda_0) < 0$, so that $D_r(\lambda_0) < 0$, exactly as wanted.

Next we show the promised monotonicity. Let's denote by $D'_r(\lambda)$ the first derivative of $D_r(\lambda)$ with respect to λ. Then we see that

$$D'_r(\lambda) = -\lambda D_{r-1}(\lambda) + \left(\frac{b_r}{a_r} - \lambda\right) D'_{r-1}(\lambda) + \frac{c_{r-1}^2}{a_{r-1}a_r} D_{r-2}(\lambda) + \frac{c_{r-1}^2}{a_{r-1}a_r} \lambda D'_{r-2}(\lambda)$$

If r is even then we argue that $D'_r(\lambda) > 0$ for all $\lambda \geq 1$. Indeed, we have that

$$-\lambda D_{r-1}(\lambda) = - \cdot + \cdot - = +$$

$$\left(\frac{b_r}{a_r} - \lambda\right) D'_{r-1}(\lambda) = - \cdot - = +$$

$$\frac{c_{r-1}^2}{a_{r-1}a_r} D_{r-2}(\lambda) = + \cdot + = +$$

$$\frac{c_{r-1}^2}{a_{r-1}a_r} \lambda D'_{r-2}(\lambda) = + \cdot + \cdot + = +$$

Since all summands of $D'_r(\lambda)$ are positive, $D_r(\lambda)$ is increasing.

Similarly, if r is odd, we show that $D'_r(\lambda) < 0$ for all $\lambda \geq 1$. Indeed, we have that

$$-\lambda D_{r-1}(\lambda) = - \cdot + \cdot + = -$$

$$\left(\frac{b_r}{a_r} - \lambda\right) D'_{r-1}(\lambda) = - \cdot + = -$$

$$\frac{c_{r-1}^2}{a_{r-1}a_r} D_{r-2}(\lambda) = + \cdot - = -$$

$$\frac{c_{r-1}^2}{a_{r-1}a_r} \lambda D'_{r-2}(\lambda) = + \cdot + \cdot - = -$$

Since all summands of $D'_r(\lambda)$ are negative, $D_r(\lambda)$ is decreasing.

In order to show that a group of strongly monotone robots converges to a stable configuration, it remains to prove all complex roots of D_r have norm < 1; this is the main idea behind the proof of Proposition 3.

Proposition 3. *For a strongly monotone collection of $n = 4$ robots the schedule produced by Algorithm 3 has the optimal idleness.*

Proof. (Proof of Proposition 3) Again, we show that the conditions of Theorem 4 are satisfied. When $n = 4$, and using (4), we can write the characteristic polynomial we need to study, that has the form $D_3(\lambda) = (A_3 - \lambda)(A_2 - \lambda)(A_1 - \lambda) + \lambda (B_2(A_3 - \lambda) + B_3(A_1 - \lambda))$, where by A_i we abbreviate b_i/a_i and by B_i

we abbreviate $c_{i-1}^2/a_{i-1}a_i$. Next we argue that all roots of D_3 have norm less than 1. By assuming strong monotonicity, Theorem 6 says that all real roots have norm between 0 and 1. Hence, we only need to check any complex roots. All we need to use below is that $0 \le A_i, B_i < 1$, and this follows by assuming simple (speed) monotonicity.

Since D_3 is of degree 3, it always has a real root, call it r, and at most two complex roots (that are conjugate to each other), say with norm $\|\rho\|$. Since the constant term of D_3 is $A_1 A_2 A_3$, it follows that $\|\rho\| = \frac{A_1 A_2 A_3}{r}$. Next we prove that $r \ge \min\{A_1, A_2, A_3\}$, concluding what we need. Indeed, consider the polynomial

$$\frac{1}{B_2 + B_3} D_3(\lambda)$$

$$= \frac{1}{B_2 + B_3}(A_3 - \lambda)(A_2 - \lambda)(A_1 - \lambda) + \lambda\left(\frac{B_2}{B_2 + B_3}(A_3 - \lambda) + \frac{B_3}{B_2 + B_3}(A_1 - \lambda)\right)$$

which clearly has the same roots as $D_3(\lambda)$. Now, the root r above is a value for λ that satisfies the following equality

$$\frac{1}{B_2 + B_3}(A_3 - \lambda)(A_2 - \lambda)(A_1 - \lambda) = -\lambda\left(\frac{B_2}{B_2 + B_3}(A_3 - \lambda) + \frac{B_3}{B_2 + B_3}(A_1 - \lambda)\right).$$

The left-hand-side polynomial, which is of degree 3, has real roots A_1, A_2, A_3, and most importantly it is decreasing for all $x \le \min\{A_1, A_2, A_3\}$ and for all $\lambda \ge \max\{A_1, A_2, A_3\}$. The right-hand-side polynomial is of degree 2, and has two real roots. One of them is 0, and the other, call it $\bar{r}$, is a convex combination of A_1, A_3, hence we have $\min\{A_1, A_3\} \le \bar{r} \le \max\{A_1, A_3\}$. Moreover, the degree 2 polynomial is negative for $0 < \lambda < \bar{r}$ and positive for $\lambda > \bar{r}$, and is increasing for all $\lambda \ge \bar{r}$. Since $\bar{r}$ is in the line segment between $\min\{A_1, A_3\}, \max\{A_1, A_3\}$, it must be the case that the graphs of the two polynomials intersect for some λ between $\min\{A_1, A_2, A_3\}$ and $\max\{A_1, A_2, A_3\}$. Therefore, $r \ge \min\{A_1, A_2, A_3\}$.

4   Conclusion

In this work we introduced the study of patrolling a unit interval using two speed robots. As it is common in the literature, we focused on the design of patrolling schedules that minimize idleness. Our findings indicate that optimal patrolling schedules can be achieved also by self-stabilizing collections of robots given that certain technical conditions hold true for their patrolling and walking speeds. This has an important implication in real-life optimization problems, e.g. in the Scheduling with Regular Delivery we introduce, since it shows that optimality can be achieved without the assumption of centrality, as well with primitive robots (i.e. robots with only minimal computation power).

A number of questions relevant to the current work remain open. Recall that we proposed some technical conditions on the robots' speeds that are sufficient for convergence of the induced dynamical systems with a limited number of robots. Are the same conditions necessary when robots start from the same endpoint of the interval? Is it possible to generalize the bouncing rule between robots that collide while moving in the same direction so that convergence is achieved using the same conditions? Are the proposed conditions (monotonicity, and strong monotonicity) sufficient for the convergence of multiple robots? Finally, the study of other patrolling domains is interesting and can be surprisingly demanding.

References

1. Agmon, N., Hazon, N., Kaminka, G.A.: The giving tree: constructing trees for efficient offline and online multi-robot coverage. Ann. Math. Artif. Intell. **52**(2–4), 143–168 (2008)
2. Agmon, N., Kraus, S., Kaminka, G.A.: Multi-robot perimeter patrol in adversarial settings. In: ICRA, pp. 2339–2345 (2008)
3. Almeida, A., Ramalho, G., Santana, H., Tedesco, P., Menezes, T., Corruble, V., Chevaleyre, Y.: Recent advances on multi-agent patrolling. In: Bazzan, A.L.C., Labidi, S. (eds.) SBIA 2004. LNCS (LNAI), vol. 3171, pp. 474–483. Springer, Heidelberg (2004). doi:10.1007/978-3-540-28645-5_48
4. Alpern, S., Morton, A., Papadaki, K.: Optimizing randomized patrols. Operational Research Group, London School of Economics and Political Science (2009)
5. Alpern, S., Morton, A., Papadaki, K.: Patrolling games. Oper. Res. **59**(5), 1246–1257 (2011)
6. Amigoni, F., Basilico, N., Gatti, N., Saporiti, A., Troiani, S.: Moving game theoretical patrolling strategies from theory to practice: an USARSim simulation. In: ICRA, pp. 426–431 (2010)
7. Angluin, D., Aspnes, J., Diamadi, Z., Fischer, M., Peralta, R.: Computation in networks of passively mobile finite-state sensors. Distrib. Comput. **18**(4), 235–253 (2006)
8. Angluin, D., Aspnes, J., Eisenstat, D., Ruppert, E.: The computational power of population protocols. Distrib. Comput. **20**(4), 279–304 (2007)
9. Bampas, E., Gąsieniec, L., Hanusse, N., Ilcinkas, D., Klasing, R., Kosowski, A.: Euler tour lock-in problem in the rotor-router model. In: Keidar, I. (ed.) DISC 2009. LNCS, vol. 5805, pp. 423–435. Springer, Heidelberg (2009). doi:10.1007/978-3-642-04355-0_44
10. Beauquier, J., Burman, J., Clement, J., Kutten, S.: On utilizing speed in networks of mobile agents. In: Proceeding of the 29th ACM SIGACT-SIGOPS Symposium on Principles of Distributed Computing, pp. 305–314. ACM (2010)
11. Chalopin, J., Das, S., Gawrychowski, P., Kosowski, A., Labourel, A., Uznański, P.: Limit behavior of the multi-agent rotor-router system. In: Moses, Y. (ed.) DISC 2015. LNCS, vol. 9363, pp. 123–139. Springer, Heidelberg (2015). doi:10.1007/978-3-662-48653-5_9
12. Chevaleyre, Y.: Theoretical analysis of the multi-agent patrolling problem. In: IAT, pp. 302–308 (2004)
13. Cieliebak, M., Flocchini, P., Prencipe, G., Santoro, N.: Distributed computing by mobile robots: gathering. SIAM J. Comput. **41**(4), 829–879 (2012)
14. Collins, A., Czyzowicz, J., Gasieniec, L., Kosowski, A., Kranakis, E., Krizanc, D., Martin, R., Morales Ponce, O.: Optimal patrolling of fragmented boundaries. In: SPAA (2013)

15. Czyzowicz, J., Gąsieniec, L., Kosowski, A., Kranakis, E.: Boundary patrolling by mobile agents with distinct maximal speeds. In: Demetrescu, C., Halldórsson, M.M. (eds.) ESA 2011. LNCS, vol. 6942, pp. 701–712. Springer, Heidelberg (2011). doi:10.1007/978-3-642-23719-5_59

16. Czyzowicz, J., Kranakis, E., Pacheco, E.: Localization for a system of colliding robots. In: Fomin, F.V., Freivalds, R., Kwiatkowska, M., Peleg, D. (eds.) ICALP 2013. LNCS, vol. 7966, pp. 508–519. Springer, Heidelberg (2013). doi:10.1007/978-3-642-39212-2_45

17. Dereniowski, D., Kosowski, A., Pajak, D., Uznanski, P.: Bounds on the cover time of parallel rotor walks. In: STACS 2014, pp. 263–275 (2014)

18. Dijkstra, E.W.: Selected Writings on Computing: A Personal Perspective. Springer, New York (1982)

19. Dumitrescu, A., Ghosh, A., Csaba, D.T.: On fence patrolling by mobile agents. CoRR, abs/1401.6070 (2014)

20. Elmaliach, Y., Agmon, N., Kaminka, G.A.: Multi-robot area patrol under frequency constraints. Ann. Math. Artif. Intell. **57**(3–4), 293–320 (2009)

21. Elmaliach, Y., Shiloni, A., Kaminka, G.A.: A realistic model of frequency-based multi-robot polyline patrolling. In: AAMAS, vol. 1, pp. 63–70 (2008)

22. Elor, Y., Bruckstein, A.M.: Autonomous multi-agent cycle based patrolling. In: Dorigo, M., Birattari, M., Caro, G.A., Doursat, R., Engelbrecht, A.P., Floreano, D., Gambardella, L.M., Groß, R., Şahin, E., Sayama, H., Stützle, T. (eds.) ANTS 2010. LNCS, vol. 6234, pp. 119–130. Springer, Heidelberg (2010). doi:10.1007/978-3-642-15461-4_11

23. Gabriely, Y., Rimon, E.: Spanning-tree based coverage of continuous areas by a mobile robot. In: ICRA, pp. 1927–1933 (2001)

24. Hare, J., Gupta, S., Wilson, J.: Decentralized smart sensor scheduling for multiple target tracking for border surveillance. In: ICRA, pp. 3265–3270. IEEE (2015)

25. Hazon, N., Kaminka, G.A.: On redundancy, efficiency, and robustness in coverage for multiple robots. Robotics Auton. Syst. **56**(12), 1102–1114 (2008)

26. Kawamura, A., Kobayashi, Y.: Fence patrolling by mobile agents with distinct speeds. In: Chao, K.-M., Hsu, T., Lee, D.-T. (eds.) ISAAC 2012. LNCS, vol. 7676, pp. 598–608. Springer, Heidelberg (2012). doi:10.1007/978-3-642-35261-4_62

27. Kosowski, A., Pajak, D.: Does adding more agents make a difference? A case study of cover time for the rotor-router. In: Esparza, J., Fraigniaud, P., Husfeldt, T., Koutsoupias, E. (eds.) ICALP 2014. LNCS, vol. 8573, pp. 544–555. Springer, Heidelberg (2014). doi:10.1007/978-3-662-43951-7_46

28. Machado, A., Ramalho, G., Zucker, J.-D., Drogoul, A.: Multi-agent patrolling: an empirical analysis of alternative architectures. In: Simão Sichman, J., Bousquet, F., Davidsson, P. (eds.) MABS 2002. LNCS (LNAI), vol. 2581, pp. 155–170. Springer, Heidelberg (2003). doi:10.1007/3-540-36483-8_11

29. Marden, M.: The Geometry of the Zeros of a Polynomial in a Complex Variable. Mathematical Surveys, vol. 3. AMS (1949)

30. Marino, A., Parker, L.E., Antonelli, G., Caccavale, F.: Behavioral control for multi-robot perimeter patrol: a finite state automata approach. In: ICRA, pp. 831–836 (2009)

31. Pasqualetti, F., Franchi, A., Bullo, F.: On optimal cooperative patrolling. In: CDC, pp. 7153–7158 (2010)

32. Yanovski, V., Wagner, I.A., Bruckstein, A.M.: A distributed ant algorithm for efficiently patrolling a network. Algorithmica **37**(3), 165–186 (2003)

Inventory Routing with Explicit Energy Consumption: A Mass-Flow Formulation and First Experimentation

Yun He[1,2(✉)], Cyril Briand[1,2], and Nicolas Jozefowiez[1,3]

[1] CNRS, LAAS, 7 avenue du colonel Roche, 31400 Toulouse, France
[2] Univ de Toulouse, UPS, LAAS, 31400 Toulouse, France
yunhe@lass.fr
[3] Univ de Toulouse, INSA, LAAS, 31400 Toulouse, France

Abstract. Energy efficiency is becoming an important criteria for the inventory systems. Our aim is to explicitly integrate the energy into the existing Inventory Routing Problem (IRP). The problem is based on a multi-period single-vehicle IRP with one depot and several customers. An energy estimation model is proposed based on vehicle dynamics. A mass-flow based Mixed Integer Linear Programming (MILP) formulation is presented. Instead of minimizing the distance or inventory cost, energy minimization is taken as an objective. Benchmark instances for inventory routing are adapted for energy estimation and experiments are conducted. The results are compared with those of the distance/ inventory cost minimization.

1 Introduction

The Inventory Routing Problem (IRP) is developed under the Vendor Managed Inventory (VMI) management model, where the supplier monitors the inventory level of each retailer and acts as a central decision maker for the long-term replenishment policy of the whole system. With respect to the traditional Retailer Managed Inventory (RMI), the VMI results in a more efficient resource utilization: on the one hand, the supplier can reduce its inventories while maintaining the same level of service, or can increase the level of service while reducing the transportation cost; on the other hand, the retailers can devote less resources to monitoring their inventories while having the guarantee that no stock-out will occur [1].

Nowadays, the inventory management is faced with a new challenge—the sustainability. As one of the three bottom lines of sustainable supply chain management, environmental sustainability is the most recognized dimension [2]. As shown in [3], energy costs account for about 60% of the total cost of a unit of cargo transported on road. Since the traditional IRP concentrates solely on the economic benefits such as transportation costs and inventory costs, there is definitely a need to study the IRP under the energy perspective.

© Springer International Publishing AG 2017
B. Vitoriano and G.H. Parlier (Eds.): ICORES 2016, CCIS 695, pp. 96–116, 2017.
DOI: 10.1007/978-3-319-53982-9_6

Under the VMI management model, the IRP combines the inventory management, vehicle routing and scheduling. There are three simultaneous decisions to make [4]:

1. when to serve a customer;
2. how much to deliver when serving a customer;
3. how to route the vehicle among the customers to be served.

These three decisions can be transformed for energy optimization.

1. The visiting time to a customer is adaptable. We can choose a delivery time that is both convenient for the customers and that can also avoid rush hours, as congestion is one of the main causes of high energy consumption and CO_2 emissions.
2. Under the VMI policy, the customer demands are flexible and can be distributed in different combinations. This property allows us to determine an optimal set of delivery quantities that is the most effective for energy use while making sure that stock-out never happens.
3. The order of visit and the vehicle routes are to be determined. It is thus possible to design a routing strategy that takes the roads with the least energy costs.

Our purpose is to explicitly incorporate energy issue into the IRP. We introduce an energy estimation method and propose a Mixed Integer Linear Programming (MILP) optimization model that integrates energy cost into the objective function. Our study concentrates on Decisions 2 and 3 presented above. We discuss the possible influence of distribution and routing strategy to the energy consumption of the inventory system. The main contributions of this paper are: (i) to propose an approach to estimate the energy consumed in the transportation activities of inventory routing; (ii) to reformulate the IRP to explicitly incorporate the energy; (iii) to analyse the possible energy savings and the trade-offs between energy savings, travelled distances and inventory costs.

The remainder of this paper is organized as follows: A brief literature review is provided in Sect. 2. Section 3 gives a description of the energy estimation method, defines the problem and presents the mathematical model. After that, experimentation and results are given in Sect. 4, followed by the conclusion in Sect. 5.

2   Literature Review

In the literature, there are a lot of studies on the IRP since its origin in the year 1980s. There are also an emerging number of papers on the environmental-related routing problems these years. However, few researchers have paid attention to the energy IRP. In the remainder of the section, we start from a general literature review of the IRPs and the Green Vehicle Routing Problems (GVRPs), then we discuss the incorporation of these two categories of problems.

2.1 Inventory Routing Problem

The IRP was first studied under the context of the distribution of industrial gases [5]. Early studies concentrate on the impact of short-term decisions to long-term inventory management and the combination of inventory management and vehicle routing [6–8]. Later on, various versions of IRP come out but there is no standard version. The IRP can be generally classified by seven criteria as shown in Table 1 [9]. For the inventory policy, under Maximum Level (ML) inventory policy, the replenishment level is flexible but bounded by the capacity available at each customer. While under Order-up-to Level (OU) policy, whenever a customer is visited, the quantity delivered is that to fill its inventory capacity. The IRP can also be considered as deterministic, stochastic or dynamic according to the availability of information on the demands.

Table 1. Classification of IRPs.

Criteria	Possible options		
Time horizon	Finite	Infinite	
Structure	One-to-one	One-to-many	Many-to-many
Routing	Direct	Multiple	Continuous
Inventory policy	Order-up-to level (OU)	Maximum level (ML)	
Inventory decisions	Lost sales	Back-order	Non-negative
Fleet composition	Homogeneous	Heterogeneous	
Fleet size	Single	Multiple	Unconstrained

Both exact and approximative methods have been studied to solve the IRP. In [1], an MILP formulation of the IRP is proposed and the first branch-and-cut algorithm is developed. [10] extended the previous formulation to cases with heterogeneous multiple vehicles, with transshipment and also with consistency constraints. They also proposed a branch-and-cut algorithm. Heuristic algorithms are widely applied in early papers, such as assignment heuristic [7], clustering heuristic [8], and trade-off based heuristic [6]. A randomized greedy algorithm is developed in [11] for the inventory routing with continuous moves which contains both pick-ups and deliveries. Later papers applied a variety of metaheuristics, such as Greedy Randomized Adaptive Search Procedure (GRASP) [12], tabu search [13], Adapted Large Neighbourhood Search (ALNS) [14], etc. Recent solution methods combine heuristic and mathematical programming, yielding the so-called "matheuristic" algorithms. For example, based on the formulation in [1], new formulations are proposed in [15]. The authors used a branch-and-cut algorithm that adds cuts heuristically and compared the new formulations with existing ones using a large set of benchmark instances.

Two literature reviews are worth mentioning here. A survey of the industrial aspects of the problem can be found in [16], and the typologies of the problem as well as their solution methods is reviewed in [9].

2.2   Green Freight Routing

In the literature, there are a growing number of papers about the green logistics and sustainable supply chain management.

On the transportation side, the eco-driving mechanism is developed to guide the driver to perform the most fuel efficient operation [17], and the eco-routing navigation systems aim to identify the most energy-efficient route for a vehicle to travel between two points in real-time [18].

More and more researchers pay attention to the incorporation of energy into Vehicle Routing Problem (VRP). In [19], the Capacitated Vehicle Routing Problem (CVRP) is extended with a new cost function that depends on both the distance travelled and the load of the vehicle and this problem is defined as the Energy Minimizing Vehicle Routing Problem (EMVRP). In [20], the Fuel Consumption Rate (FCR), a factor depending on load, is added to the CVRP with the objective of minimizing fuel consumption. The authors proposed a String-model-based simulated annealing algorithm to solve the problem. They discovered that the difference of the FCR induced by vehicle loads, the diverse demands and uneven geographical positions of the customers can all influence fuel cost savings. Focusing on the pollution and CO_2 emission generated by the road transport sector, the Pollution Routing Problem (PRP) is proposed to explicitly control the Greenhouse Gas (GHG) emission of the transportation [21]. In this paper, the authors discussed the trade-offs between distance, load and energy minimizing objectives and the influence of parameters such as vehicle speed and load as well as customers time windows.

A detailed literature review of the GVRP can be found in [22]. In this review, the environmental sensitive Vehicle Routing Problem is divided into three groups: the Green-VRP for the optimization of energy consumption [19,20]; the PRP for the reduction of pollution, especially GHG emissions [21]; and the Vehicle Routing in Reverse Logistics for the collection of wastes and end-of-life product. It is pointed out that incorporating inventory models with PRP models can be promising.

The vehicle emission models and their applications to road transportation planning are reviewed in [23]. Different factors affecting fuel consumption or vehicle emission are discussed in this paper. However, few models focus on the energy consumed.

The first paper that incorporates environmental aspects in the IRP is a case study from the petrochemical industry [24]. They proposed a mathematical model that integrate CO_2 cost into the objective function. A multi-vehicle inventory routing with CO_2 emission is also studied in [25], where a penalty is added if the CO_2 emission of the planning route is higher than a predefined capacity of the road. However, neither of them consider the energy estimation, nor are the influential factors clearly exposed.

3 Problem Statement

The problem in our study is based on a *multi-period single-vehicle deterministic* IRP with *one depot and several customers*. The planning horizon is defined by a set of periods. The vehicle can leave the depot only once per period. In each period, it makes a tour around the customers that need to be refilled and returns to the depot. Stock-out and back-orders are not allowed in this model. Instead of the distance and inventory minimization, we take energy minimization as objective. Both the ML and the OU policy are applied to see the influence of different replenishment strategies to the energy consumption.

In this section, we start with the energy estimation model that explains how to estimate the energy cost of a road segment with vehicle speed variation. Then we present the parameters and decisions variables of the problem. Finally we give the mathematical formulation of the problem.

3.1 Energy Estimation Model

In most of the literature related to IRP, transportation cost is represented by the distance travelled. This is not suitable for the energy minimization because energy is influenced by various factors as shown in Sect. 2.2. In the literature related to energy issue, most models focus on the fuel consumption or CO_2 emission and most of them depend on vehicle type. Nevertheless, with the emergence of electric and hybrid vehicles, we find it more appropriate to estimate the energy used directly. In addition, for the generality of the problem, it is important that the energy estimation model would apply for every type of vehicles.

According to [26], travelling kinematic variation (accelerations, idle duration, etc.) obviously affects engine load and by turns the energy consumption. Thus, in this paper, we propose a general simple model based on vehicle dynamics. This model would be applicable to European suburban transportation network with short or medium distances and potentially high traffic intensity. It can give us a gross estimation of the energy required by a vehicle on a road segment with speed variation, independent of vehicle type or energy source.

General Parameters. Suppose a vehicle travelling from one location to another. The path of the vehicle between two locations is supposed to be predefined with an average stop rate τ, and the total distance travelled is s. So the vehicle stops $\tau \cdot s$ times during the trip. The coefficient of friction is a fixed parameter $\mu = 0.01$. The gravitational acceleration is $g = 9.81 m/s^2$. The environmental effects of the road (wind, temperature etc.) as well as the viscosity of air are ignored. Road slopes, denoted by the rate θ, is deduced from the difference of altitude between the origin and the destination ($\sin\theta = \frac{\Delta h}{s}$ where $\Delta h = h_{end} - h_{begin}$). We consider that on the same segment of road, the vehicle only climb or descend once. If the vehicle climbs (the destination is higher than the origin), then θ is positive. If the vehicle descends, then θ is negative. The only forces exerted on the vehicle are the gravity, the rolling resistance and the traction force of the engine.

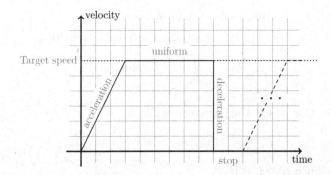

Fig. 1. The speed variation of the vehicle with time.

In our model, the stop rate τ, i.e., the number of stops per unit of distance is used to model the dynamics of the vehicle on a fixed segment of road. This parameter can also represent the traffic condition on the road. More precisely, with a traffic near free flow, τ takes a value near 0, which means that the vehicle goes through the road fluently without any stops; however, with congestion, this number corresponds to a higher value to indicate a frequent speed variation. Usually τ takes a value between 0 and 4 depending on road types [27]. Moreover, there exists an interrelationship between the distance travelled, the stop rate and the speed and acceleration of the vehicle, which is generally explained in Sect. 4.1.

Between every two stops, the vehicle speed is supposed to follow a fixed pattern of variation—acceleration, uniform speed movement and stop. Each time, the vehicle speeds up from 0 to the target speed V with a fixed acceleration a_{acc}. It goes on at this speed for a while and then stops. The stop is supposed to be instantaneous. This pattern is repeated $\tau \cdot s$ times supposing that the vehicle has no speed at both the starting and the ending point. After each stop, it speeds up again to the same target speed. This speed profile is shown in Fig. 1.

Energy Consumption Formula. According to knowledge of physics and energy conservation, under the hypothesis of speed variation presented above, the energy consumption on a segment between two locations with a distance s can be calculated as follows:

Acceleration Phase. In this phase, the speed of the vehicle increases from 0 to the target speed V with a constant acceleration a_{acc}. If we note F_{acc} the traction force of the engine, s_{acc} the distance travelled on the slope θ, E_{acc} the energy consumed, and $P_{\text{acc}}(t)$ the engine power at instant t, we have:

$$v(t) = a_{\text{acc}}t;$$
$$V = a_{\text{acc}}t_{\text{acc}};$$
$$s_{\text{acc}}(t) = \frac{1}{2}a_{\text{acc}}t^2;$$

$$F_{\text{acc}} - mg\mu\cos\theta - mg\sin\theta = ma_{\text{acc}};$$
$$P_{\text{acc}}(t) = F_{\text{acc}}v(t) = m(a_{\text{acc}} + g\mu\cos\theta + g\sin\theta)a_{\text{acc}}t;$$
$$E_{\text{acc}} = \int_0^{t_{\text{acc}}} P_{\text{acc}}(t)dt.$$

At the end of this phase, the engine power is

$$P_{\text{acc}}^{\text{end}} = m(a_{\text{acc}} + g\mu\cos\theta + g\sin\theta)V;$$

the total distance travelled is

$$s_{\text{acc}} = \frac{V^2}{2a_{\text{acc}}}.$$

The total energy cost per unit of mass is:

$$c_{\text{acc}} = \frac{1}{2}(a_{\text{acc}} + g\mu\cos\theta + g\sin\theta)a_{\text{acc}}t_{\text{acc}}^2$$
$$= \frac{1}{2}V^2 + g(\mu\cos\theta + \sin\theta)\frac{V^2}{2a_{\text{acc}}}.$$

Uniform-Speed Phase. In this phase, the vehicle travels in uniform speed for a distance s_u. This distance is computed as the difference between the total distance s and the total distance travelled in acceleration and deceleration. Since the deceleration is considered to be instantaneous ($s_{\text{dec}} = 0$), the total distance travelled at uniform speed is calculated as:

$$s_u = s - \tau\, s\,(s_{\text{acc}} + s_{\text{dec}}) = s - \tau\, s\, s_{\text{acc}}$$

with $\tau \cdot s$ the total number of stops. The engine force is constant

$$F_u = \mu mg\cos\theta + mg\sin\theta = mg(\mu\cos\theta + \sin\theta).$$

The engine power is also constant

$$P_u(t) = F_u V = mg\mu V.$$

The total energy cost per unit of mass in the uniform phase is:

$$c_u = g(\mu\cos\theta + \sin\theta)(s - \tau\, s\, s_{\text{acc}}).$$

Deceleration Phase. In this phase, since we consider an instantaneous stop, the distance s_{dec} and the engine power P_{dec} is 0, the energy is lost immediately: $E_{\text{dec}} = \frac{1}{2}mV^2$. The energy cost per unit of mass is

$$c_{\text{dec}} = \frac{1}{2}V^2.$$

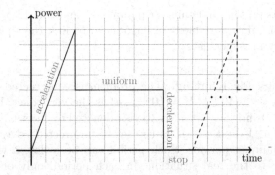

Fig. 2. The power variation with time.

Summary of Energy Cost. According to the three-phase analysis of the vehicle energy, the total energy cost per unit of mass when distance s is travelled with stop rate τ is then:

$$c = c_u + \tau \cdot s \left(c_{\text{acc}} + c_{\text{dec}}\right).$$

Finally, we get:

$$c = g(\mu \cos\theta + \sin\theta)s + \tau s V^2$$

Since θ is usually a small value, we can take $\cos\theta \simeq 1$ and $\sin = \frac{\Delta h}{s}$, then the energy cost would be

$$c = g\mu s + g\Delta h + \tau s V^2, \tag{1}$$

which is the same result as obtained by the law of conservation of energy.

Figure 2 shows the power variation of the vehicle under the previous speed variation. We can see that each time the vehicle speeds up, there appears a "peak" of engine power which corresponds to a potentially high energy consumption. This is also reflected by (1)—the more the vehicle stops on a road segment (τ takes a bigger value), the higher the energy would cost. In addition, if the vehicle climbs a mountain (Δh positive), more energy would be used.

In this way, we define $c_{ij} = g\mu s_{ij} + g\Delta h_{ij} + \tau_{ij} s_{ij} V_{ij}^2$ the energy cost per unit of mass from location i to location j. It is related to the distance travelled s_{ij}, the variation of altitude Δh_{ij} and the dynamics of the vehicle on the road as expressed by the stop rate τ_{ij} and target speed V_{ij}.

3.2   Problem Definition

To facilitate the energy estimation, two units are used to measure inventory— the number of components and the weight in kilograms (kg). The number of components is used by the customers to represent their inventory levels and to count the number of packages of delivered goods. The weight is used by the transporters. It is the physical mass of the components transported by the vehicle.

The next parts present in details the parameters and variables of the problem. In particular, mass flow variables are introduced to link the energy estimation and the inventory management.

General Settings for Routing. The problem is constructed on a *complete undirected* graph $\mathcal{G} = \{V, E\}$. $V = \{0, \ldots, n\}$ is the vertex set. It includes one depot denoted by 0 and the customers to visit denoted by the set $V_c = \{1, \ldots, n\}$. $E = \{(i, j) \mid i, j \in V \text{ and } i < j\}$ is the set of undirected edges. There are T replenishment planning periods. Each period can be a day, a week or even a month according to the real situation. In each period, only one tour can be performed. If a tour is presented in a period, the vehicle starts from the depot, travels through all the customers who need to be served at this period and returns back to the depot at the end of the period.

Three sets of decision variables z_i^t, x_{ij}^t and y_{ij}^t correspond to routing. For each $i \in V_c, t \in T$, z_i^t is a binary variable indicating whether customer i is served at period t. It equals 1 if customer i is served and 0 otherwise. Particularly, z_0^t indicates whether the tour at period t is performed (equals 1) or not (0). For each edge $(i, j) \in E$ and each period $t \in T$, x_{ij}^t is an integer variable indicating the number of times that edge (i, j) is used in the tour of period t. $x_{ij}^t \in \{0, 1\}$ if $i, j \in V_c$ since a customer can be visited only once per period and $x_{0j} \in \{0, 1, 2\} \ \forall j \in V_c$ because direct shipping is allowed between the depot and a customer. For each arc $(i, j) \in V \times V$ and each period $t \in T$, variable y_{ij}^t is a binary variable to indicate the direction of the vehicle route. It equals 1 if the vehicle travels from i to j at period t.

The vehicle has a capacity Q expressed in numbers of components and a mass limit M. The empty vehicle mass, or curb weight of the vehicle is W (in kg).

Inventory Characteristics. Inventory levels at customers and depot are monitored during the whole planning time horizon. They are summarised at the end of each replenishment period. The customer demands are described as demand rates per period. In each period, r_i is the number of units of components demanded by the customer $i \in V_c$. In particular, r_0 is the number of components made available at the depot in each period. Each customer $i \in V_c$ has a stocking capacity C_i, while the depot is supposed to have an unlimited stocking capacity. h_i is the inventory storage cost per unit of component per period at customer i or the depot.

Two variables are defined for the inventory management. The variable I_i^t is the inventory level in number of components at the depot 0 or at the customer $i \in V_c$ at the end of period t. The variable q_i^t is the number of components delivered to customer $i \in V_c$ during period $t \in T$.

Commodity Mass Flow. The energy cost per unit of mass for each arc $(i, j) \in V \times V$ is defined at the end of Sect. 3.1. If m_{ij} is the mass (kg) loaded on the vehicle when traversing from vertex i to vertex j and W the vehicle weight(kg), the energy cost of the vehicle travelling from i to j is thus:

$$c_{ij}(m_{ij} + W) \tag{2}$$

As we can see from (2), the total energy cost is a linear function of mass. Meanwhile, the mass or the quantity of products is also an important element in the inventory management. It is a measurement of the inventory levels. In fact, there exists a mass flow inside the transportation network and it can serve as a bridge linking the inventory routing and the energy optimization.

In the traditional IRP formulations presented in [15], a flow formulation exists to model the inventory flows inside the transportation network. Our model takes advantage of this formulation. Instead of thinking the flows in terms of number of components, the mass of the shipped components is considered. In each period, once we decide the mass transported on each edge of the network, we can deduce the number of components left at each customer vertex. Or inversely, if we know how many units of components are delivered to each customer, we can decide the order of visits and get a mass flow in the transportation network that minimizes the energy consumed.

In our model, variables m_{ij}^t are defined as the mass transported by the vehicle from i to j at period t. They are linked with the vehicle flow variables y_{ij}^t. If the vehicle does not go from i to j at period t ($y_{ij}^t = 0$), m_{ij}^t is equal to 0.

Figure 3 details the various flows traversing customer i at period t. The inventory flow I_i^t and the demand r_i, expressed in number of components, are associated with the dotted arcs. They describe the variation of the inventory level of i with time periods. The solid arcs stand for the mass of the incoming and outgoing products (m_{ji}^t and m_{ij}^t respectively). They are used to estimate the potential energy consumption, with c_{ij}^t the energy cost per unit of mass on edge (i,j). The weight of one unit of component in kilograms at a customer $i \in V_c$ is denoted by m_i. The difference $\dfrac{1}{m_i}\Big(\sum\limits_{j \in V \setminus \{i\}} m_{ji}^t - \sum\limits_{j \in V \setminus \{i\}} m_{ij}^t\Big)$ gives the number of components q_i^t delivered to customer i during period t.

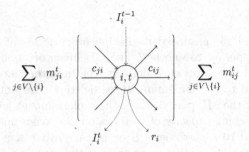

Fig. 3. The flows passing through customer i at period t.

3.3   Mathematical Model

With the parameters and the variables defined in Sect. 3, the mathematical model is explained here. The complete model can be found in the annexe.

Objectives. Two objectives are defined, one for inventory and distance optimization and the other for energy optimization. Equation (3) is the traditional objective as defined in [15]. It is the sum of the total distance travelled plus the sum of the inventory storage costs over all the periods.

$$\min \sum_{t \in T} \sum_{(i,j) \in V \times V} s_{ij} y_{ij}^t + \sum_{t \in T} \sum_{i \in V} h_i I_i^t \tag{3}$$

Equation (4) is the sum of the total energy consumed in the inventory routing over all the periods. Note that it contains two terms: one is a flexible cost related to the transported mass of the vehicle m_{ij}^t, and the other is a fixed cost induced by the vehicle curb weight W.

$$\min \sum_{t \in T} \sum_{(i,j) \in V \times V} c_{ij} m_{ij}^t + W \sum_{t \in T} \sum_{(i,j) \in V \times V} c_{ij} y_{ij}^t \tag{4}$$

Constraints. Compared with the basic flow formulation in [15], mass flow variables take place of commodity flow variables.

Inventory Management. Constraints (5) to (9) are for monitoring the inventory levels of each location at each period.

$$I_0^t = I_0^{t-1} + r_0 - \sum_{i \in V_c} q_i^t \qquad \forall t \in T \tag{5}$$

$$I_i^t = I_i^{t-1} - r_i + q_i^t \qquad \forall i \in V_c, t \in T \tag{6}$$

$$q_i^t \geq C_i z_i^t - I_i^{t-1} \qquad \forall i \in V_c, t \in T \tag{7}$$

$$q_i^t \leq C_i - I_i^{t-1} \qquad \forall i \in V_c, t \in T \tag{8}$$

$$q_i^t \leq C_i z_i^t \qquad \forall i \in V_c, t \in T \tag{9}$$

Constraints (5) and (6) ensure that the inventory levels of each station are coherent from one period to another. The OU inventory policy is ensured by constraints (7) and (8)—after each delivery, the inventory level of each visited customer is fulfilled to the maximum. If we delete Constraints (7), the model becomes one under the ML policy, where the replenishment level is flexible but bounded by the stocking capacity of each customer. Constraints (9) ensure that nothing is delivered to a customer i if he is not visited at a period and that otherwise, the delivered quantity never exceeds the capacity.

Commodity Mass Flow Management. Constraints (10) and (11) are the mass flow constraints.

$$\sum_{j \in V_c} m_{0j}^t = \sum_{i \in V_c} q_i^t m_i \qquad \forall t \in T \qquad (10)$$

$$\sum_{j \in V} m_{ji}^t - \sum_{j \in V} m_{ij}^t = q_i^t m_i \qquad \forall i \in V_c, t \in T \qquad (11)$$

Constraints (10) ensure that at period t, the mass out of the depot is equal to the total mass transported to all the customers. Constraints (11) ensure that for each customer i at each period t, the quantity received is equal to the difference between the entering and the leaving mass flow.

Vehicle Routing. Constraints (12) to (17) are typical routing constraints.

Degree Constraints

$$\sum_{j \in V_c} x_{0j}^t = 2z_0^t \qquad \forall t \in T \qquad (12)$$

$$\sum_{\substack{j \in V \\ j < i}} x_{ji}^t + \sum_{\substack{j \in V_c \\ j > i}} x_{ij}^t = 2z_i^t \qquad \forall i \in V_c, t \in T \qquad (13)$$

Directed Vehicle Flow

$$\sum_{j \in V_c} y_{0j}^t = z_0^t \qquad \forall t \in T \qquad (14)$$

$$\sum_{j \in V} y_{ij}^t = z_i^t \qquad \forall t \in T, i \in V_c \qquad (15)$$

$$\sum_{j \in V} y_{ji}^t = z_i^t \qquad \forall t \in T, i \in V_c \qquad (16)$$

$$x_{ij}^t = y_{ij}^t + y_{ji}^t \qquad \forall t \in T, (i,j) \in E \qquad (17)$$

Constraints (12) and (13) are the degree constraints. They define the route of the vehicle in each period. Constraints (14)–(16) restrain the direction of the vehicle flow. They link y and z variables to make sure that in each period at most one tour is performed and that each customer is visited at most once in each period. Constraints (17) link variables y and x to ensure that each edge is used at most once in each period.

Vehicle Capacity. Constraints (18) and (19) guarantee that the vehicle capacity is never exceeded both in number of components and in unit of mass.

$$\sum_{i \in V_c} q_i^t \le Q z_0^t \qquad \forall t \in T \qquad (18)$$

$$m_{ij}^t \le M y_{ij}^t \qquad \forall t \in T, (i,j) \in V \times V \qquad (19)$$

Constraints (19) also link the mass flow and the vehicle flow on the graph. They make sure that the direction of the vehicle flow is the same as that of the mass flow.

Variable Domains. Constraints (20)–(26) are the variable domains.

$$0 \le I_i^t \le C_i, I_i^t \in \mathbb{N} \qquad\qquad \forall i \in V, t \in T \tag{20}$$

$$0 \le q_i^t \le Q, q_i^t \in \mathbb{N} \qquad\qquad \forall i \in V_c, t \in T \tag{21}$$

$$0 \le m_{ij}^t \le M, m_{ij}^t \in \mathbb{N} \qquad\qquad \forall (i,j) \in V \times V, t \in T \tag{22}$$

$$z_i^t \in \{0,1\} \qquad\qquad \forall i \in V, t \in T \tag{23}$$

$$x_{ij}^t \in \{0,1\} \qquad\qquad \forall (i,j) \in E, i < j, t \in T \tag{24}$$

$$x_{0j}^t \in \{0,1,2\} \qquad\qquad \forall j \in V_c, t \in T \tag{25}$$

$$y_{ij}^t \in \{0,1\} \qquad\qquad \forall (i,j) \in V \times V, t \in T \tag{26}$$

All the variables take integer values. Constraints (20) ensure that the inventory level of a customer never exceeds his stocking capacity. Constraints (21) and (22) make sure that the vehicle capacity is never exceeded neither in terms of mass nor in terms of units of components. Note that for variables x_{0j}^t, since direct routing is possible, they can be assigned with value 2.

4 Experimentation and Results

The existing IRP instances proposed in [1] are adapted for energy estimation. The MILP model is constructed and solved using the adapted instances. An analysis of the obtained results is presented.

4.1 Data Generation

Information on stop rates τ and vehicle target speeds V relative to the distance is added to the benchmark instances proposed in [1]. The correlation within these parameters is determined based on empirical data of delivery trucks on real routes provided by [28]. The following part explains how the data set is generated.

First, two types of road is considered—highway and national route. For each edge between two locations, the type of road is generated randomly. The target speed and the number of stops for different types of roads are generated using different methods. On a highway, the maximum speed is fixed at 110 km/h, and the number of stops is fixed at 2 stops per edge no matter how long is travelled. On a national route, the vehicle speed is fixed at 80 km/h and the number of stops is linearly dependent on the distance with a random error. For all types of road, the average acceleration rate is fixed at 1.01 m/s^2. The instances generated contain two categories of type proportion: one is with $\frac{2}{3}$ edges among all the edges defined as highway and $\frac{1}{3}$ as national route; the other is with $\frac{1}{3}$ edges among all defined as highway and $\frac{2}{3}$ as national route.

Then, a random number between 1 and 10 is generated for each customer i to represent the mass of one unit of components m_i. Vehicle weight and mass capacity are correlated according to vehicle information provided in [29].

Last, a random number between 0 to 500 is generated as the altitude h_i of each location i.

In total, 64 cases are generated. Each case contains 5 instances. The cases are categorized by the number of periods (3 or 6 periods of replenishment planning), the proportion of the inventory storage cost in relation to the transportation cost (high or low), the inventory replenishment policy (OU or ML), the proportion of each type of road in the whole map and the number of customers in the map.

4.2 System Settings

The model is realized in C++ with $IBM^{®}$ $ILOG^{®}$ $CPLEX$ $12.6.1.0$ and solved by the default Branch-and-Bound algorithm with one thread. The operation system is $Ubuntu$ 14.04 LTS with $Intel^{®}$ $Core^{®}$ $i7\text{-}4790$ $3.60\,GHz$ processor and $16\,GB$ memory.

The solution process is divided into two phases. In the first phase, the objective is to minimize the combined cost of transportation and inventory as in objective function (3). In the second phase, starting with the solution of the first phase, the same model is solved to minimize the total energy consumption as computed in objective function (4).

A time limit of $1800\,s$ is set for each of the two phases. All the other settings of CPLEX are as default. The results of both of the two phases are compared in the next part.

4.3 Result and Analysis

Performance. The dimension of an instance is determined by the number of periods and the number of customers. The inventory policy (OU or ML) changes the constraint sets of the model. The combination of these three parameters define a category of instances. Each category contains 20 instances. In Table 2, computation time in seconds of each solution phase ("time1" and "time2") and the solution status within the time limit ("status1" and "status2") are listed for each category. The values for computation time are average values over all the instances of the same category. If all the instances of a category can be solved to optimality by CPLEX, the status is noted "Optimal". If part of the instances of a category can be solved to optimality, then the status is noted "Optimal(n)" with a number n in parentheses indicating the number of instances solved to optimality in this category. Otherwise, if no optimal solution is found in the time limit by CPLEX, then the average relative gap after $1800\,s$ of computation is reported as the status, and the time value is noted 1800.

As we can see from Table 2, energy minimization is much more difficult to solve than inventory and transportation cost minimization (time2 $\gg$ time1). This may result from the large possible combination of the values of the mass flows. The problem is NP since it is an extension of the VRP. It becomes more difficult as the dimension of the instances increases. For both OU and ML policies, instances larger than 20 customers with 3 periods or 15 customers with 6 periods can hardly be solved to optimality for energy optimization within the time limit. The influence of the inventory policy to energy minimization is not as obvious as in the traditional IRP.

Table 2. Solution status and solving time.

T	n	ML policy				n	OU policy			
		status1	status2	time1	time2		status1	status2	time1	time2
3	5	Optimal	Optimal	0.137	0.0892	5	Optimal	Optimal	0.0992	0.0844
	10	Optimal	Optimal	1.78	1.60	10	Optimal	Optimal	1.70	1.34
	15	Optimal	Optimal	12.5	61.5	15	Optimal	Optimal	16.6	35.0
	20	Optimal	Optimal(13)	199	976	20	Optimal	Optimal(17)	65.6	749
	25	Optimal	0.084	67.01	1800	25	Optimal(12)	0.058	787	1800
	30	Optimal	0.12	310	1800	30	Optimal(12)	0.10	1017	1800
	35	Optimal	0.15	183	1800	35	Optimal(8)	0.15	1248	1800
	40	Optimal(16)	0.17	624	1800	40	Optimal(2)	0.19	1714	1800
	45	Optimal(14)	0.18	756	1800	45	0.054	0.22	1800	1800
	50	Optimal(5)	0.23	1649	1800	50	0.10	0.27	1800	1800
6	5	Optimal	Optimal	2.53	0.401	5	Optimal	Optimal	0.489	0.465
	10	Optimal	Optimal	45.0	54.9	10	Optimal	Optimal	29.2	55.3
	15	Optimal	Optimal(1)	429	1790	15	Optimal	Optimal(3)	169	1630
	20	Optimal(4)	0.098	1639	1800	20	Optimal(7)	0.10	1487	1800
	25	Optimal(4)	0.14	1575	1800	25	Optimal(6)	0.16	1515	1800
	30	0.077	0.20	1800	1800	30	0.075	0.21	1800	1800

Energy Impacting Factors. Suppose that the energy consumption in Phase 1 is noted E_1 and the consumption in Phase 2 is noted E_2. The energy reduction in the following paragraphs is defined as the ratio $r = \frac{E_2 - E_1}{E_1}$. In general, the energy reduction can achieve 35% in average. It is at least 21% and can reach as high as 46%.

Several factors have an impact on the energy reduction. First, the size of the instance can influence the potential energy reduction. Larger instances tend to induce higher energy conservation. Figure 4 shows the variation of the energy reduction in relation with the number of customers.

Second, there exists a compromise between the inventory/distance cost and energy cost since all the energy reduction necessitates an augmentation of distance and inventory costs whatever policy or planning horizon (see Fig. 5). And it seems that under the configuration where there are more national routes (that means a more variation of number of stops), this compromise becomes more obvious.

Third, energy reduction does not mean distance minimization. Contrary to the common belief that the shortest route is the one that minimizes the energy, our study shows that both the distance and the vehicle weight and loads are important for the energy consumption. The vehicle with a high load tends to start his journey with the least energy cost road and put to the end the visit to a customer in an area with high energy cost. For example, Fig. 6 shows the route of the vehicle under different objectives with 3 periods and OU policy. Figure 6(a) is the route obtained with energy minimization. The vehicle serves Customer 1

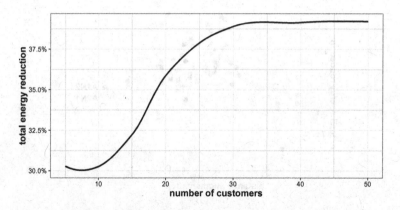

Fig. 4. Number of customers and energy reduction.

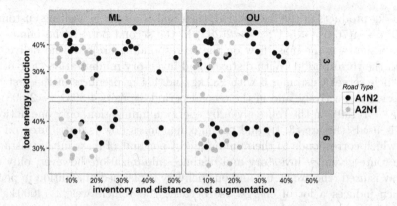

Fig. 5. Distance and inventory cost and energy cost under different configurations.

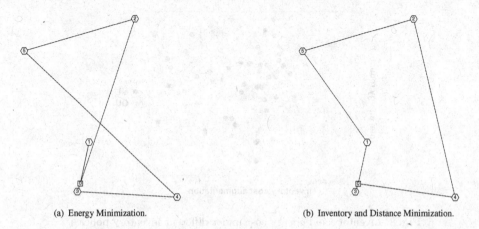

(a) Energy Minimization.   (b) Inventory and Distance Minimization.

Fig. 6. Vehicle routes under different objectives.

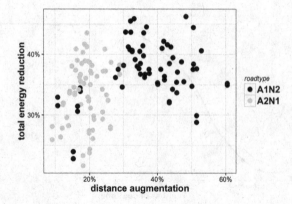

Fig. 7. Distance and energy cost under different road types.

with 65 kg products in the first period, then makes a tour by visiting customers $3(1230) \rightarrow 4(766) \rightarrow 5(478) \rightarrow 2(280)$ in the second period (the number in the parentheses is the mass flow on the corresponding arc), and no delivery is done in the third period. With distance and inventory minimization (Fig. 6(b)), the vehicle serves Customer 3 with 232 kg and 464 kg products in the first and last period respectively, and in the second period, it visits $4(896) \rightarrow 2(608) \rightarrow 5(328) \rightarrow 1(130)$. In the route given by energy minimization, only one national route is used (the arc $(3,4)$) and the maximal mass flow is distributed on arc $(0,3)$ which corresponds to the minimum cost per unit of mass in this instance. In the route given by inventory and distance minimization, however, only one highway is used (the arc $(1,0)$) and one delivery is planned in addition in period 3, which induces a lot of energy use because the vehicle weight (4000 kg) is important in relation to the payload (464 kg).

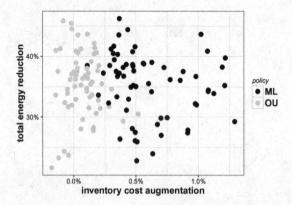

Fig. 8. Inventory and energy cost under different inventory policies.

The relation between the distance and the energy is influenced by the road type (or vehicle dynamics such as number of stops and vehicle speed). As shown in Fig. 7, in a world with more highways (road type "A2N1" means that 2/3 arcs of all the arcs are highway and 1/3 are national route, similar for "A1N2"), we can achieve 40% of energy reduction with at most 30% of augmentation of distance, whereas in "A1N2" configuration, the augmentation of distance can be as high as 60% to have an energy reduction of 35%. This confirms the fact that a free-flow configuration is better for energy use.

Last but not least, inventory replenishment strategy can also impact the energy reduction potential of an inventory routing system (See Fig. 8). Under ML policy, inventory change to save energy is higher than under OU policy, since ML policy is more flexible than OU policy.

5   Conclusions

Energy consumption is an important aspect in both economical and ecological view. It becomes more and more important with the sustainable requirement of the inventory systems. We address the combination of inventory management, vehicle routing and energy minimization and propose a new mass-flow based formulation of the IRP with explicit energy consumption. This formulation uses an energy estimation methods depending on vehicle dynamics (speed) and road characteristics (stop rate per kilometre and slope). This estimation gives us an energy cost function that is linear to the total mass. In this formulation, the mass is added as a decision variable and the energy cost function is considered as an objective. Our first experimentation shows that there is a great potential in improving the energy efficiency in the inventory routing.

Various parameters can have an impact on the energy consumption. From the transportation aspect, vehicle speed and number of stops are important. The improvement of energy can be higher on a road with congestion. On the inventory management side, inventory strategy can influence the energy consumption. Under the condition that no customer is in stock-out, the ML policy provides much more flexibility for energy minimization than the OU policy.

Further works include modelling of traffic networks, so that different traffic conditions as well as vehicle speed levels could be considered in the decision process. More data are needed from the real world to accomplish this work.

The inventory routing model needs to be improved to better control the time and quantity of each delivery. In fact, the traditional IRP is very aggregated in terms of inventory levels and delivery time since all the inventory monitoring is summarized in periods. In reality, however, there are two time scales for the routing and inventory management. It is in small scale (minutes or hours) for the vehicle routing because traffic conditions can change in a day, while for the inventory management it is in large scale (days or months) according to the real application. The model should incorporate these two time scales so that there is no loss of information.

Heuristics are being studied to speed up the computation. Especially with realistic data, larger number of customers or longer decision periods is common.

The extension of the problem to a multi-objective one is also a promising track of study.

Acknowledgements. This work was supported by the ECO-INNOVERA-1rst call EASY (ANR-12-INOV-0002).

Appendix

The energy minimizing IRP mathematical model with OU inventory policy is presented below.

$$\min \sum_{t \in T} \sum_{(i,j) \in V \times V} c_{ij} m_{ij}^t + W \sum_{t \in T} \sum_{(i,j) \in V \times V} c_{ij} y_{ij}^t$$

$$I_0^t = I_0^{t-1} + r_0 - \sum_{i \in V_c} q_i^t \qquad \forall t \in T$$

$$I_i^t = I_i^{t-1} - r_i + q_i^t \qquad \forall i \in V_c, t \in T$$

$$q_i^t \geq C_i z_i^t - I_i^{t-1} \qquad \forall i \in V_c, t \in T$$

$$q_i^t \leq C_i - I_i^{t-1} \qquad \forall i \in V_c, t \in T$$

$$q_i^t \leq C_i z_i^t \qquad \forall i \in V_c, t \in T$$

$$\sum_{j \in V_c} m_{0j}^t = \sum_{i \in V_c} q_i^t m_i \qquad \forall t \in T$$

$$\sum_{j \in V} m_{ji}^t - \sum_{j \in V} m_{ij}^t = q_i^t m_i \qquad \forall i \in V_c, t \in T$$

$$\sum_{j \in V_c} x_{0j}^t = 2 z_0^t \qquad \forall t \in T$$

$$\sum_{\substack{j \in V \\ j < i}} x_{ji}^t + \sum_{\substack{j \in V_c \\ j > i}} x_{ij}^t = 2 z_i^t \qquad \forall i \in V_c, t \in T$$

$$\sum_{j \in V_c} y_{0j}^t = z_0^t \qquad \forall t \in T$$

$$\sum_{j \in V} y_{ij}^t = z_i^t \qquad \forall t \in T, i \in V_c$$

$$\sum_{j \in V} y_{ji}^t = z_i^t \qquad \forall t \in T, i \in V_c$$

$$x_{ij}^t = y_{ij}^t + y_{ji}^t \qquad \forall t \in T, (i,j) \in E$$

$$\sum_{i \in V_c} q_i^t \leq Q z_0^t \qquad \forall t \in T$$

$$m_{ij}^t \leq M y_{ij}^t \qquad \forall t \in T, (i,j) \in V \times V$$

$$0 \leq I_i^t \leq C_i, I_i^t \in \mathbb{N} \qquad \forall i \in V, t \in T$$

$$0 \leq q_i^t \leq Q, q_i^t \in \mathbb{N} \qquad \forall i \in V_c, t \in T$$
$$0 \leq m_{ij}^t \leq M, m_{ij}^t \in \mathbb{N} \qquad \forall (i,j) \in V \times V, t \in T$$
$$z_i^t \in \{0,1\} \qquad \forall i \in V, t \in T$$
$$x_{ij}^t \in \{0,1\} \qquad \forall (i,j) \in E, i < j, t \in T$$
$$x_{0j}^t \in \{0,1,2\} \qquad \forall j \in V_c, t \in T$$
$$y_{ij}^t \in \{0,1\} \qquad \forall (i,j) \in V \times V, t \in T$$

References

1. Archetti, C., Bertazzi, L., Laporte, G., Grazia Speranza, M.: A branch-and-cut algorithm for a vendor-managed inventory-routing problem. Transp. Sci. **41**, 382–391 (2007)
2. Fish, L.A.: Applications of Contemporary Management Approaches in Supply Chains. InTech (2015)
3. Sahin, B., Yilmaz, H., Ust, Y., Guneri, A.F., Gulsun, B.: An approach for analysing transportation costs and a case study. Eur. J. Oper. Res. **193**, 1–11 (2009)
4. Campbell, A.M., Clarke, L.W., Savelsbergh, M.W.P.: 12.inventory routing in practice. In: Toth, P., Vigo, D. (eds.) The Vehicle Routing Problem, pp. 309–330. Society for Industrial and Applied Mathematics, Philadelphia (2001)
5. Bell, W.J., Dalberton, L.M., Fisher, M.L., Greenfield, A.J., Jaikumar, R., Kedia, P., Mack, R.G., Prutzman, P.J.: Improving the distribution of industrial gases with and on-line computerized routing and scheduling optimizer. Interfaces **13**, 4–23 (1983)
6. Burns, L.D., Hall, R.W., Blumenfeld, D.E., Daganzo, C.F.: Distribution Strategies that Minimize Transportation and Inventory Costs (1985)
7. Dror, M., Ball, M.: Inventory/routing: reduction from an annual to a short-period problem. Naval Res. Log. (NRL) **34**, 891–905 (1987)
8. Anily, S., Federgruen, A.: One warehouse multiple retailer systems with vehicle routing costs. Manag. Sci. **36**, 92–114 (1990)
9. Coelho, L.C., Cordeau, J.F., Laporte, G.: Thirty years of inventory routing. Transp. Sci. **48**, 1–19 (2013)
10. Coelho, L.C., Laporte, G.: The exact solution of several classes of inventory-routing problems. Comput. Oper. Res. **40**, 558–565 (2013)
11. Savelsbergh, M., Song, J.H.: Inventory routing with continuous moves. Comput. Oper. Res. **34**, 1744–1763 (2007)
12. Campbell, A.M., Savelsbergh, M.W.P.: A decomposition approach for the inventory-routing problem. Transp. Sci. **38**, 488–502 (2004)
13. Archetti, C., Bertazzi, L., Hertz, A., Grazia Speranza, M.: A hybrid heuristic for an inventory routing problem. INFORMS J. Comput. **24**, 101–116 (2012)
14. Coelho, L.C., Cordeau, J.F., Laporte, G.: The inventory-routing problem with transshipment. Comput. Oper. Res. **39**, 2537–2548 (2012)
15. Archetti, C., Bianchessi, N., Irnich, S., Grazia Speranza, M.: Formulations for an inventory routing problem. Int. Trans. Oper. Res. **21**, 353–374 (2014)
16. Andersson, H., Hoff, A., Christiansen, M., Hasle, G., Løkketangen, A.: Industrial aspects and literature survey: combined inventory management and routing. Comput. Oper. Res. **37**, 1515–1536 (2010)

17. Alam, M.S., McNabola, A.: A critical review and assessment of Eco-Driving policy & technology: benefits & limitations. Transp. Policy **35**, 42–49 (2014)

18. Boriboonsomsin, K., Barth, M.J., Zhu, W., Vu, A.: Eco-routing navigation system based on multisource historical and real-time traffic information. IEEE Trans. Intell. Transp. Syst. **13**, 1694–1704 (2012)

19. Kara, İ., Kara, B.Y., Yetis, M.K.: Energy minimizing vehicle routing problem. In: Dress, A., Xu, Y., Zhu, B. (eds.) COCOA 2007. LNCS, vol. 4616, pp. 62–71. Springer, Heidelberg (2007). doi:10.1007/978-3-540-73556-4_9

20. Xiao, Y., Zhao, Q., Kaku, I., Xu, Y.: Development of a fuel consumption optimization model for the capacitated vehicle routing problem. Comput. Oper. Res. **39**, 1419–1431 (2012)

21. Bektaş, T., Laporte, G.: The pollution-routing problem. Transp. Res. Part B: Methodol. **45**, 1232–1250 (2011)

22. Lin, C., Choy, K.L., Ho, G.T.S., Chung, S.H., Lam, H.Y.: Survey of green vehicle routing problem: past and future trends. Expert Syst. Appl. **41**, 1118–1138 (2014)

23. Demir, E., Bektaş, T., Laporte, G.: A review of recent research on green road freight transportation. Eur. J. Oper. Res. **237**, 775–793 (2014)

24. Treitl, S., Nolz, P.C., Jammernegg, W.: Incorporating environmental aspects in an inventory routing problem. A case study from the petrochemical industry. Flex. Serv. Manuf. J. **26**, 143–169 (2014)

25. Alkawaleet, N., Hsieh, Y.F., Wang, Y.: Inventory routing problem with CO2 emissions consideration. In: Operations, L. (ed.) Supply Chain Management and Sustainability, pp. 611–619. Cham, Springer, Berlin (2014)

26. Samaras, Z., Ntziachristos, L.: Average hot emission factors for passenger cars and light duty trucks. Technical report LAT report No. 9811, Lab. of Applied Thermodynamics, Aristotle University of Thessaloniki (1998)

27. André, M., Hassel, D., Weber, F.J.: Development of short driving cycles-Short driving cycles for the inspection of in-use cars – Representative European driving cycles for the assessment of the I/M schemes. Technical report May, INRETS - LEN, Laboratoire Énergie Nuisances (1998)

28. Walkowicz, K., Duran, A., Burton, E.: Fleet DNA project data summary report (2014). Accessed 20 Mar 2015

29. EcoTransIT World Initiative (EWI): Ecological transport information tool for worldwide transports methodology and data—update. Technical report, IFEU Heidelberg and INFRAS Berne and IVE Hannover (2014)

Delineation of Rectangular Management Zones and Crop Planning Under Uncertainty in the Soil Properties

Víctor M. Albornoz[✉], José L. Sáez, and Marcelo I. Véliz

Departamento de Industrias, Universidad Técnica Federico María,
Campus Santiago Vitacura, Av. Santa María, 6400 Santiago, Chile
victor.albornoz@usm.cl, {jose.saezt,marcelo.veliz}@alumnos.usm.cl

Abstract. In this article we cover two problems that often farmers have to face. The first one is to generate a partition of an agricultural field into rectangular and homogeneous management zones according to a given soil property, which has variability in time that is presented by a set of possible scenarios. The second problem assigns the correct crop rotation for those management zones defined before. These problems combine aspects of precision agriculture and optimization with the purpose of achieving a site and time specific management of the field that is consistent and effective in time for a medium term horizon. Thus, we propose a two-stage stochastic integer programming model with recourse that solves the delineation problem facing a finite number of possible scenarios, after this we propose a deterministic crop planning model, and then we combine them into a new two-stage stochastic program that can solve both problems under ucertainty conditions simultaneously. We describe the proposed methodology and the results achieved in this research.

Keywords: OR in agriculture · Stochastic programming · Management zones · Crop planning · Precision agriculture

1 Introduction

In agriculture, spatial variability of the soil properties is a key aspect in yield and quality of crops. In fact, one of the problems in precision agriculture consists in dividing the field into site specific management zones, which based on a soil property such as: pH, organic matter, phosphorus, nitrogen, crop yield, etc. Delineating rectangular zones into zones relatively homogeneous allows better agricultural machines performance and eases the design of irrigation systems, being also important to consider the zones size and the total amount of management zones from the field partition.

The problem of defining management zones in presence of site specific variability has been studied in [6], where an integer programming model for determining rectangular zones is defined, this problem considers spatial variability of an specific soil property and choose the best field partition. The main idea

© Springer International Publishing AG 2017
B. Vitoriano and G.H. Parlier (Eds.): ICORES 2016, CCIS 695, pp. 117–131, 2017.
DOI: 10.1007/978-3-319-53982-9_7

is to define homogeneous management zones to optimize the use of inputs for crops. The model is solved by the complete enumeration of the variables, thus it is possible only to solve small and medium size instances due to the problem is NP-hard. To deal with this problem, a column generation algorithm was proposed in [3] which allows to efficiently solve large instances of the problem.

Recently, this previous problem has been applied for irrigation systems design, see [9] where linear programming is used as one of the methods for delineating management zones. Other methods for delineation are classified as clustering methods, see [12, 13, 17], but their major drawback is the resulting fragmentation of the zones, because these methods generate oval shaped and disjoint zones.

On the other hand, there is an important problem related to choose the correct crop planning in these management zones previously defined. This problem has been studied in [2], where a hierarchical scheme is shown, the first step is to define de rectangular zones and the second step is to define the correct crop to be cultivated in each zone with a single-objective that maximize profits. Other authors have studied this crop planning problem with multi-objective models like in [16] where the objective is to maximize profits and minimize a monthly irrigation planning. Within this context, the crop rotation problem rises, in these kind of models the objective is to find an optimal set of crops for a temporal horizon, it is also possible to change crops in each period. This problem is studied in [7] where a linear problem is developed to define the optimal crop rotation plan subject to certain ecologically-based constraints and considering that harvested crops can be stocked but only for a limited period of time.

Although the problem of defining management zones in presence of site specific variability has been studied in previous works, to the best of our knowledge, an important characteristic that has not been considered yet is the variability in time of the chosen soil property. Based on cited works, first we propose a two-stage stochastic programming model with recourse that solves field partition problem considering the chosen soil property as a random variable which can be modeled by a finite number of scenarios. Then we extend the previous model by combining the field partition problem with the crop planning problem into a new two-stage stochastic program. This is a new proposal based on the fact that the crop yield depends directly of the soil propierties where it is cultived.

Stochastic programming is chosen in these situations because deterministic models are not capable of adding the effect of uncertainty to the solutions. Stochastic programming is based on considering random variables that are described by a number of possible scenarios; see e.g. [4, 19, 20].

In the last few years, stochastic programming has been used more often in a wide variety of applications due to its capacity of solving problems increasingly large, thus more realistic models, see e.g. [8, 22] for general applications.

In agriculture, stochastic programming has been used to solve different problems related with situations where uncertainty is a key aspect in the decision making process. Besides delineation decision there are other important decisions to make, as crop planning, water planning, food supply chain and agricultural raw materials supply planning, among others. Crop planning is a decision where a crop pattern must be chosen for each management zone, this pattern last a specific number of crop cycles and thus must face future weather scenarios and prices, see [11,14,24]. Water planning is important because the need for more agricultural production requires large amounts of water for irrigation purposes, making water resources scarce, thus surface water resources must be allocated among farmers and also a plan must be made for the use of this water, see [5] and [15]. Stochastic programming is also applied in agricultural supply chain problems, as food supply chain where a growing and distribution plan must be made, and raw materials supply where a raw material acquisition plan must be made considering that some raw materials are seasonal, in these problems variability appears in the form of weather conditions and product demands, see [1,23].

Within stochastic programming models exists the two-stage models with recourse. These models recognize two types of decisions that must be made sequentially. First stage decision or here-and-now must be made previously to the performance of the random variables. Then, second stage decision or wait-and-see, which must compensate the effects of the first stage decisions once the performance of the random variables are known, due to this, the variables in this stage are denoted as recourse variables. The goal of these models consists in finding the optimal first stage decision that minimize total costs, defined by the sum of the first stage decision costs and the expected costs of the second stage decisions; see e.g. [10].

For example, in the proposed model for generating a partition, the first stage decision chooses a field partition that minimizes the number of management zones; these zones must satisfy certain homogeneity level that depends on the performance value of the sample points which are the random variables in this case. On the other hand, second stage decision uses looseness variables that relax homogeneity constraints in exchange of a penalty. This penalty helps to achieve management zones homogeneity goal while minimizes the use of the looseness variables.

In this article, problem formulation needs the generation of the total number of potential management zones; in other words, problem solving considers the complete enumeration of zones is known. This is feasible for small and medium size instances as the ones used in this work, which represents a good starting point to approach to this problem. Although, proposed formulation can be extended to large instances by the application of a column generation algorithm, but its use exceeds the purpose of the present research, see [3].

In following sections, the article is organized as follows. Section 2 details the proposed models to solve the delineation problem, the crop planning problem and the combined problem, from data collection to the solving process itself. After this, in Sect. 3, results obtained by the application of proposed methodology are presented. Finally in Sect. 4, main conclusions and future works from the application of the model are presented.

2   Materials and Methods

As we mentioned before, this work consists in generating a field partition composed by a group of management zones based on a chosen soil property which has variability in space and time, and also choosing the optimal crop plan that minimizes the cost involved in the production horizon period. The proposed methodology has two steps. First, the task is to model the soil property space variability by taking samples on the field, this process must be done several times in different periods to measure variability in time, with this data, instances are generated. And the second step, consists on solving the problems using two different approaches: The first one consists in a hierarchical scheme where we define a two-stage stochastic integer programming model for the delineation problem that minimizes the number of management zones in its first stage and minimizes noncompliance of the homogeneity level in the second stage, and a deterministic crop planning model that selects the correct crop pattern that should be cultivated in the previously defined management zones in a certain period of time. The second approach consists in defining a combined two-stage stochastic model that covers both problems with the same uncertainty condition that were considered in the stochastic delineation model.

2.1   Instance Generation

In the first step, we generate instances that will be solved by the models. To achieve this is necessary to use specialized software as MapInfo; this software creates thematic maps of the field that summarizes and shows spatial variability of the soil properties measured from the sample points. This includes sample coordinates, pH level, organic matter index, phosphorus, base sum, crop yield, etc. As an example, Fig. 1 shows two thematic maps from the same field, one with organic matter (OM) and the other with phosphorus (P). In OM case, green zones represent reference levels of OM, while sky blue and blue zones represent zones with 34.8% and 3.97% above normal values of OM, also red and yellow zones presents values with 6.06% and 3.97% under normal OM values. On the other hand, in P case sky blue and blue zones are 27.31% and 10.34% above normal, and red and yellow zones are 13.79% and 48.27%, respectively. Both maps show spatial variability of these indexes in a field, this proves the importance of dividing the field into management zones with uniform characteristics, to apply inputs needed in each zone through site specific farming.

Also, we need to include variability in time of the measured indexes. For that, we use thematic map data sets from the same field for several time periods; these

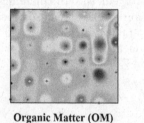
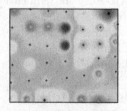

Organic Matter (OM)   **Phosphorus (P)**

Fig. 1. Organic matter and phosphorus map. (Color figure online)

will be used either to generate the probability distribution function of the soil property or to create different scenarios with each one of these instances. A possible value of the random variable consist in assign a specific value to each of the sample points on the field, i.e., the random variable is represented by a vector that includes each one of the sample points; this vector has a finite number of possible values. Scenario probabilities are assigned depending on the number of instances and the time between each sampling process. It is important to notice that a field partition is a medium term decision, i.e., this partition will last a specific number of years and after that horizon is reached, another partition must be set, thus the model must take into account possible changes in soil properties during this time. This article uses only historical data for scenario creation, but it is also valid to consider forecasts for future periods in the scenario creation step, but this exceeds the purpose of this article.

Finally, potential management zones are generated (Z set) through an algorithm that uses all sample points (S set) as inputs. As an example, in Fig. 2 there is an instance with 42 sample point field (6 rows and 7 columns) and three potential management zones from a total of 588, each one of them has rectangular form and includes at least one sample point.

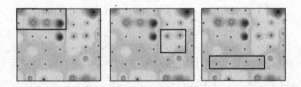

Fig. 2. Potential management zones example.

A relationship matrix $C = (c_{sz})$ is created from potential zones generation, where $c_{sz} = 1$ means that potential zone z includes sample point s, and $c_{sz} = 0$ otherwise, for every $z \in Z$, $s \in S$. Besides, index variance $\sigma^2_{z\omega}$ is obtained for each potential quarter z and each scenario $\omega \in \Omega$, where Ω is the set of possible scenarios. Both parameters are used in the model presented in the following section.

2.2 Optimization Model for the Delineation Problem

Proposed model consist in a two-stage stochastic integer programming model with recourse. In the first stage, the problem minimizes the number of management zones that cover the entire field. In the second stage, the problem minimizes noncompliance of the homogeneity level using looseness variables for each scenario but with a penalty cost for using them. This second stage is necessary because field partition must be chosen before knowing random variables performance, and it must satisfy the homogeneity constraint for any scenario, this is achieved by minimizing the expected value of the penalty for the noncompliance of the homogeneity level.

Sets, parameters and variables used in the model are described below:
   Sets:
Z: set of potential management zones, with $z \in Z$.
S: set of sample points of the field, with $s \in S$.
Ω: set of possible scenarios, with $\omega \in \Omega$.

   Parameters:
c_{sz}: Coefficient that represents if quarter z covers sample point s or not.
M_ω: Penalty cost per unit for noncompliance of the required homogeneity level.
n_z: Number of sample points in quarter or management zone z.
p_w: Probability of scenario ω.
$\sigma_{z\omega}^2$: Quarter variance z calculated from the soil property in scenario ω.
$\sigma_{T\omega}^2$: Total variance of the field calculated from the soil property data in scenario ω.
N: Total number of sample points.
UB: Upper bound for the number of management zones chosen.
α: Required homogeneity level.

   Decision variables:

$$q_z = \begin{cases} 1, & \text{if quarter } z \text{ is assigned to field partition} \\ 0, & \text{otherwise, } z \in Z \end{cases}$$

h_ω: Looseness for the homogeneity level in scenario $\omega \in \Omega$.

The two-stage stochastic model with recourse is presented now:

$$Min \; \sum_{z \in Z} q_z + \sum_{\omega \in \Omega} p_\omega Q(q, h_\omega) \tag{1}$$

$$s.t.$$

$$\sum_{z \in Z} c_{sz} q_z = 1 \quad \forall s \in S \tag{2}$$

$$\sum_{z \in Z} q_z \leqslant UB \tag{3}$$

$$q_z \in \{0, 1\} \quad \forall z \in Z \tag{4}$$

$$Where \; Q(q, h_\omega) = Min M_\omega h_\omega \tag{5}$$

$$s.t.$$

$$h_\omega \geqslant \sum_{z \in Z} [(n_z - k)\sigma_{z\omega}^2 + (1 - \alpha)\sigma_{T\omega}^2] q_z - (1 - \alpha)\sigma_{T\omega}^2 N \tag{6}$$

$$h_\omega \geqslant 0 \tag{7}$$

Problems (1)–(4) correspond to the first stage decision, while (5)–(7) correspond to the second stage decision. Objective function (1) minimizes the sum of management zones chosen and minimizes the expected value of the penalty cost for noncompliance of the required homogeneity level, these are first and second stage objective functions respectively. Constraint (2) is typical for set partition models, guarantees that each sample point on the field is assigned only to one quarter. Constraints (3) establishes an upper bound to the number of management zones chosen to divide the field. Constraint (4) defines that quarter variables must be binary. Objective function (5) represents second stage decision for each scenario. Constraint (6) states that a required homogeneity level must be accomplished; this constraint is made from the linear version of the relative variance concept and a looseness variable for each scenario. Finally, constraint (7) states nature of second stage variables.

It is important to notice that this model, as in [6], uses an equivalent linear version of the constraint related to the relative variance concept. However in this case, as we have different possible scenarios, we must meet homogeneity level in each one of these scenarios, thus we will have a relative variance constraint for each scenario. As we have to choose only one field partition we need a way to deal with uncertainty because otherwise we will have to choose the best field partition for worst possible scenario in terms of relative variance. We propose to add new variables named as looseness variables as part of the second stage decision to get a solution that considers all possible scenarios, meeting the required homogeneity level in each one of these, and without being forced to solve the problem for the worst scenario.

Constraint (6) is created from the following non-linear constraint used in [6]:

$$1 - \frac{\sum_{z \in Z}(n_z - k)\sigma_z^2 q_z}{\sigma_T^2[N - \sum_{z \in Z} q_z]} \geqslant \alpha \tag{8}$$

This constraint uses relative variance concept, presented in [18], is a widely used criteria to measure effectiveness of chosen management zones and it must be equal or higher to a given α value, which is the required homogeneity level, that should be at least 0.5 to validate an ANOVA test hypothesis assuming k degrees of freedom. To create constraint (6) first we need to linearize Eq. (8) obtaining the following expression:

$$(1 - \alpha)\sigma_T^2[N - \sum_{z \in Z} q_z] \geqslant \sum_{z \in Z}(n_z - k)\sigma_z^2 q_z \tag{9}$$

Then if we reorder Eq. (9) we obtain:

$$\sum_{z \in Z}[(n_z - k)\sigma_z^2 + (1 - \alpha)\sigma_T^2]q_z \leqslant (1 - \alpha)\sigma_T^2 N \tag{10}$$

As we have a number of possible scenarios we define a relative variance constraint for each one of these, and also different parameters for each scenario ω:

$$\sum_{z \in Z}[(n_z - k)\sigma_{z\omega}^2 + (1 - \alpha)\sigma_{T\omega}^2]q_z \leqslant (1 - \alpha)\sigma_{T\omega}^2 N \tag{11}$$

Here is when we add the looseness variables h_ω to the right side of Eq. (11):

$$\sum_{z \in Z}[(n_z - k)\sigma_{z\omega}^2 + (1 - \alpha)\sigma_{T\omega}^2]q_z \leqslant (1 - \alpha)\sigma_{T\omega}^2 N + h_\omega \tag{12}$$

These variables allow the problem to choose a field partition that considers all possible scenarios and meet all relative variance constraints by relaxing the right side of Eq. (11) for each scenario, thus finally obtaining constraint (6). It is important to notice that looseness variables are added to the linear version of this constraint to have only linear constraints in the model.

2.3   Optimization Model for the Crop Planning Problem

The model presented in the previous section is capable of determine the optimal delineation of the field considering variability in time, however, we are going to cover a second important problem, as was mentioned before, it consists in define what kind of crop should be cultivated for a period of time. This problem is known as Crop Planning Problem. The proposed model is based on (Santos et al. 2011), this model guarantees that the demand for a specific period of time is supplied, defining a limit for the area to be used.

The sets, parameters and variables are described below:

Sets:

K: Set of potential crop rotation plans, with $k \in K$.
T: Set of periods of time, with $t \in T$.
I: Set of crops, with $i \in I$.

Parameters:

L_{kz}: Cost for cultivating the rotation k in the management zone
D_{it}: Demand for the crop i in the period t
A_{it}^k: Amount of crop i harvested in period t in crop rotation plan k.

Variable:

$$x_{kz} = \begin{cases} 1, & \text{if quarter z with crop rotation k is assigned to field partition} \\ 0, & \text{otherwise}, k \in K, z \in Z \end{cases}$$

The new model is as follows:

$$Min \sum_{z \in Z} \sum_{k \in K} L_{kz} x_{kz} \tag{13}$$

$$s.t.$$

$$\sum_{z \in Z} \sum_{k \in K} c_{sz} x_{kz} = 1 \quad \forall s \in S \tag{14}$$

$$\sum_{z \in Z} \sum_{k \in K} x_{kz} \leqslant UB \tag{15}$$

$$\sum_{z \in Z} \sum_{k \in K} A_{it}^k x_{kz} \geqslant D_{it}, \quad \forall i \in I, t \in T \tag{16}$$

$$\sum_{z \in Z} \sum_{k \in K} [(n_z - k)\sigma_z^2 + (1-\alpha)\sigma_T^2] x_{kz} \leqslant (1-\alpha)\sigma_T^2 N \tag{17}$$

$$x_{kz} \in \{0,1\} \quad \forall z \in Z, k \in K \tag{18}$$

Objective function (13) minimizes the cost of cultivate the crop rotation k in management zones chosen. Constraint (14) guarantees that each sample point on the field is assigned only to one quarter and one crop rotation plan. Constraint (15) establishes an upper bound to the number of management zones chosen to divide the field. Constraint (16) ensures that a specific demand in each period must be achieved for each crop. Contraint (17) assumes a minimum level for the homogeneity level. Constraint (18) defines that management zone variables with a crop rotation plan assigned must be binary.

2.4   Optimization Model for Field Delineation and Crop Planning Problem

The field delineation problem was defined in the model of the Sect. 2.2, and the deterministic crop rotation problem was proposed in the Sect. 2.3. With these models is possible to get an answer to these two important problems separately in a hierarchical approach. The second approach considers a new two-stage stochastic programming model that determines simultaneously what is the best field delineation and what to cultivate there, all of this under uncertainty conditions due to the variability in time of the chosen soil property. Thus, using the previous notation, we propose the combined two-stage stochastic program. The proposed model is as follows:

$$Min \ \sum_{z \in Z} \sum_{k \in K} L_{kz} x_{kz} + \sum_{w \in \Omega} p_w Q(x, h_w) \tag{19}$$

$$s.t.$$

$$\sum_{z \in Z} \sum_{k \in K} c_{sz} x_{kz} = 1 \quad \forall s \in S \tag{20}$$

$$\sum_{z \in Z} \sum_{k \in K} x_{kz} \leqslant UB \tag{21}$$

$$\sum_{z \in Z} \sum_{k \in K} A_{it}^k x_{kz} \geqslant D_{it}, \quad \forall i \in I, t \in T \tag{22}$$

$$x_{kz} \in \{0, 1\} \quad \forall z \in Z, k \in K \tag{23}$$

$$Where \ Q(x, h_w) = Min M_w h_w \tag{24}$$

$$s.t.$$

$$h_w \geqslant \sum_{z \in Z} \sum_{k \in K} [(n_z - k)\sigma_{zw}^2 + (1 - \alpha)\sigma_{Tw}^2] x_{kz} - (1 - \alpha)\sigma_{Tw}^2 N \tag{25}$$

$$h_w \geqslant 0 \tag{26}$$

Problems (19)–(23) corresponds to the first stage model, while (24)–(26) are the second stage model. Objective function (19) minimizes the cost of cultivate the crop rotation k in management zones chosen and minimizes the expected value of the penalty cost for noncompliance of the required homogeneity level, these are first and second stage objective functions respectively. Constraints

(20)–(23) are the same as constraints (14)–(17). Objective function (24) represents second stage decision for each scenario. Constraint (25)–(26) states the same condition presented in constraints (6)–(7).

3 Results

To analyze the delineation and crop planning models behavior we used one instance for the problem, using crop yield as soil property because this index has strong variability in time. In this instance, there are six possible scenarios, all of them with similar probabilities, where the two latest scenarios are more likely to occur. Chosen parameter values are:

$$M_\omega = 1.5 \quad \forall \omega \in \Omega$$
$$UB = 40$$
$$\alpha = 0.9$$
$$p_\omega = 0.15 \quad \omega \in \{1, ..., 4\}$$
$$p_\omega = 0.2 \quad \omega \in \{5, 6\}$$

We worked with one instance with 42 sample points, that generates 588 potential management zones. The number of potential management zones is obtained by the formula $\frac{((n+1)n(m+1)m)}{4}$ presented in [3], where $n = 6$ is the number of sample points in length and $m = 7$ is the number of sample points in width. The cultivating cost is estimated according to data sheets from analysis in the south of Chile. The rest of the parameters are calculated from crop yield data for each scenario.

The different models were solved with a Lenovo Thinkpad with processor Intel Core i3-2310M 2.10 GHz with 4 Gb RAM memory by using AMPL and CPLEX 12.4.

3.1 Instance Solving

In this Section, the results obtained from this instance are compared using the two approaches described before.

In the hierarchical approach, first we solved two-stage model (1)–(7). The optimal solution can be seen in Fig. 3.

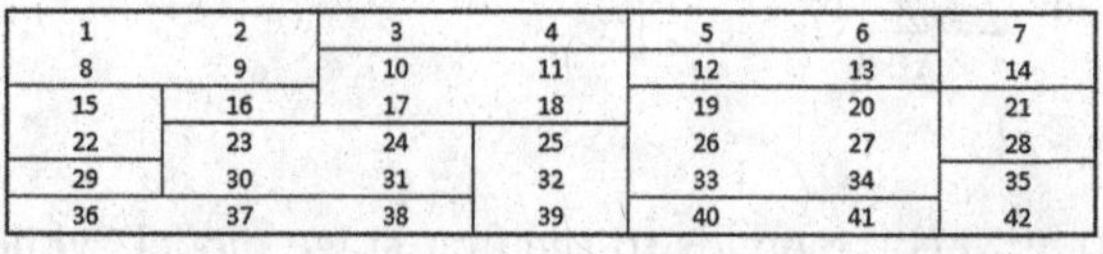

1	2	3	4	5	6	7
8	9	10	11	12	13	14
15	16	17	18	19	20	21
22	23	24	25	26	27	28
29	30	31	32	33	34	35
36	37	38	39	40	41	42

Fig. 3. Optimal solution for stochastic delineation model.

Figure 3 shows the entire field and every rectangle represents a single rectangular management zone. The numbers on it are the sample points. For this instance, the optimal solution is a plot with only 16 management zones and a penalty value of 1.83.

Once the delineation problem is solved, the farmer can choose which crop rotation plan is the best choice to cultivate, using the model (13)–(18), this is how a hierachical approach works. The results are shown as follows in Fig. 4.

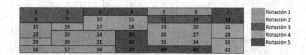

Fig. 4. Crop planning decision over optimal solution for stochastic model.

Now, it would be interesting to compare this solution with the result that can be obtained after applying the combined two-stage stochastic model (19)–(26). That model chooses a different delineation with another crop rotation plan in each management zone, this can be seen in Fig. 5.

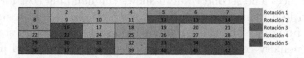

Fig. 5. Optimal solution for simultaneous model.

In order to compare with the previous results, this problem can also be solved for its average scenario by using a deterministic hierarchical approach that consists in obtaining a field delineation for the average scenario and then evaluate this solution with the stochastic model so we can measure the penalty cost for the noncompliance of the required homogeneity level in each one of the scenarios, after this the crop planning model is used to obtain the crop rotation plan for this field delineation.

The results for the delineation problem are shown in Fig. 6.

1	2	3	4	5	6	7
8	9	10	11	12	13	14
15	16	17	18	19	20	21
22	23	24	25	26	27	28
29	30	31	32	33	34	35
36	37	38	39	40	41	42

Fig. 6. Optimal solution for average scenario.

In this case, the optimal solution considers less management zones than the previous results, creating a field partition with only seven management zones

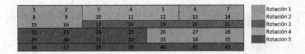

Fig. 7. Crop Planning decision over optimal solution for average model.

Table 1. Results from combined model and hierarchical models.

Instances models	Handling cost	Penalty	O.F.
Hierarchical stochastic approach	340	1,831	341,831
Combined stochastic approach	37	14.562	51.562
Deterministic hierarchical approach	116	71,558	187,558

with a penalty cost equal to 0,61. Once the delineation problem is solved, we obtain an optimal crop planning using model (13)–(18), this is shown in Fig. 7. Finally the results are summarized in Table 1.

This table shows 4 columns, the first column shows the specific approach used to run the instance. The second column shows the handling cost for cultivating a rotation plan in a management zone, in other words, this represents the first stage decision function. The third column represents the second stage decision related to the penalty cost for the noncompliance of the required homogeneity level using looseness variables for each scenario. Finally, the fourth columns shows the total value for the objective function.

In this table it is importante to notice that costs and penalties in the objective function have a high increase when we use a hierarchical approach, reaching an increase over 200%. Also, there are two importants results: the first one is that the best choice for reducing total cost is the combined stochastic model, and the second one is that the best approach that reduces the noncompliance penalty cost is the hierarchical approach using the stochastic delineation decision. This table also shows that the worst result related to the homogeneity requirements is obtained by the hierarchical deterministic approach, this is because the average solution doesn't consider the homogeneity requirements of each scenario separately. In this instance was better to cultivate in the seven management zones, proposed for average scenario delineation, than in the seventeen management zones proposed for the stochastic model, this is because the handling cost increase has a higher impact on the objective function than the penalty cost increase. This could change depending on the parameter values chosen. However the best option is to solve both problems with the combined two-stage stochastic model, this model shows significantly better results than the other approaches. For this reason, it is important to make a decision that considers the field delineation and the crop rotation plan simultaneously. Thus, with this model is possible to achieve a good solution that ensures high savings and that will improve handling work.

4 Conclusions and Further Research

This work presents two different approaches, the first one is hierarchical approach that considers a two-stage stochastic integer programming model to solve the field delineation problem facing uncertainty conditions represented by a soil property and a deterministic crop planning model with the objective to solve what was the best option to cultivate in the management zones defined before. The second approach consists in a combined two-stage stochastic model that minimizes the cost related to cultivate a specific crop rotation at the same time that chooses the field delineation of management zone, also considering the variability in time of the soil property. In Both approaches, models solutions define an optimal field partition with a recourse function that considers looseness variables that help to achieve the required homogeneity level. These approaches were applied to a real instance with 42 sample points, and it showed that the combined stochastic approach is a better choice than a hierarchical approach. The combined model minimizes handling costs related to cultivate a specific crop rotation plan in each management zone, under uncertainy condition of the soil properties, and the results were at least 200% more cheap than the hierarchical approaches.

This methodology covers small size instance solving by the complete enumeration of all potential management zones, this also needs computation of parameters described in Sect. 2.1 for each potential management zone. This is not feasible for large instances due to the problem is NP-hard and the number of variables increases at an exponential rate when number of sample points grows, thus we need more computational effort to calculate all the parameters for each variable. To deal with this issue, we propose to design a decomposition method for the combined two-stage stochastic model based in column generation to solve large instances without using all the problem variables. This will be developed based on the decomposition of the deterministic version of the delineation model presented in this article, see [3], because structure is similar, and management zones can be added as columns in the algorithm as well. Also, it would be interesting to include uncertainty in the handling costs for the crop rotation decision.

Acknowledgements. This research was partially supported by Dirección General de Investigación, Innovación y Postgrado (DGIIP) from Universidad Técnica Federico Santa María, Grant USM 28.15.20. José Luis Sáez and Marcelo Véliz wish to acknowledge the Graduate Scholarship also from DGIIP.

References

1. Ahumada, O., Villalobos, J.R., Mason, A.N.: Tactical planning of the production and distribution of fresh agricultural products under uncertainty. Agric. Syst. **112**, 17–26 (2012)
2. Albornoz, V.M., Cid-García, N.M., Ortega, R., Ríos-Solis, Y.A.: A hierarchical planning scheme based on precision agriculture. In: Plà-Aragones, L.M. (ed.) Handbook of Operational Research in Agriculture and the Agri-Food Industry, pp. 129–162. Springer, Heidelberg (2015)
3. Albornoz, V.M., Nanco, L.J.: An empirical design of a column generation algorithm applied to a management zone delineation problem. In: Fonseca, R.J., Weber, G.-W., Telhada, J. (eds.) Computational Management Science. LNEMS, vol. 682, pp. 201–208. Springer, Heidelberg (2016). doi:10.1007/978-3-319-20430-7_26
4. Birge, J., Loveaux, F.: Introduction to Stochastic Programming, 2nd edn. Springer, New York (2011)
5. Bravo, M., González, I.: Applying stochastic goal programming: a case study on water use planning. Eur. J. Oper. Res. **196**, 1123–1129 (2009)
6. Cid-García, N.M., Albornoz, V.M., Ortega, R., Ríos-Solis, Y.A.: Rectangular shape management zone delineation using integer linear programming. Comput. Electron. Agric. **93**, 1–9 (2013)
7. Costa, M.A., Dos Santos, L., Alem, D., Santos, R.: Sustainable vegetable crop supply problem with perishable stocks. Ann. Oper. Res. **219**, 265–283 (2011)
8. Gassmann, H.I., Ziemba, W.T.: Stochastic Programming. Applications in Finance, Energy, Planning and Logistics. World Scientific Publishing Company, Singapore (2012)
9. Haghverdi, A., Leib, B.G., Washington-Allen, R.A., Ayers, P.D., Buschermohle, M.J.: Perspectives on delineating management zones for variable rate irrigation. Comput. Electron. Agric. **117**, 154–167 (2015)
10. Higle, J.L.: Stochastic programming: optimization when uncertainty matters. In: Tutorials in Operations Research. INFORMS, New Orleans (2005)
11. Itoh, T., Ishii, H., Nanseki, T.: A model of crop planning under uncertainty in agricultural management. Int. J. Prod. Econ. **81–82**, 555–558 (2003)
12. Jaynes, D., Colvin, T., Kaspar, T.: Identifying potential soybean management zones from multi-year yield data. Comput. Electron. Agric. **46**(1), 309–327 (2005)
13. Jiang, Q., Fu, Q., Wang, Z.: Study on delineation of irrigation management zones based on management zone analyst software. In: Li, D., Liu, Y., Chen, Y. (eds.) CCTA 2010. IAICT, vol. 346, pp. 419–427. Springer, Heidelberg (2011). doi:10.1007/978-3-642-18354-6_50
14. Li, M., Guo, P.: A coupled random fuzzy two-stage programming model for crop area optimization-a case study of the middle Heihe River Basin, China. Agric. Water Manag. **155**, 53–66 (2015)
15. Liu, J., Li, Y.P., Huang, G.H., Zeng, X.T.: A dual-interval fixed-mix stochastic programming method for water resources management under uncertainty. Resour. Conserv. Recycl. **88**, 50–66 (2014)
16. Mainuddin, M., Das Grupta, A., Raj Onta, P.: Optimal crop planning model for an existing groundwater irrigation project in Thailand. Agric. Water Manag. **33**, 43–62 (1996)
17. Ortega, J.A., Foster, W., Ortega, R.: Definition of sub-stands for precision forestry: an application of the fuzzy k-means method. Ciencia e Investigación Agraria **29**(1), 35–44 (2002)

18. Ortega, R., Santibañez, O.A.: Determination of management zones in corn (Zea mays L.) based on soil fertility. Comput. Electron. Agric. **58**, 49–59 (2007)
19. Ramos, A., Alonso-Ayuso, A., Perez, G.: Optimizacion bajo incertidumbre. Publicaciones de la Universidad Pontificia Comillas, España, Biblioteca Comillas (2008)
20. Ruszczynski, A., Shapiro, A.: Stochastic programming. In: Handbooks in Operations Research and Management Science, vol. 10. New York, North-Holland (2003)
21. Santos, L.M.R., Costa, A.M., Arenales, M.N., Santos, R.H.S.: Sustainable vegetable crop supply problem. Eur. J. Oper. Res. **204**, 639–647 (2010)
22. Wallace, S. W., Ziemba, W. T. Applications of Stochastic Programming. MOS-SIAM Series on Optimization (2005)
23. Wiedenmann, S., Geldermann, J.: Supply planning for processors of agricultural raw materials. Eur. J. Oper. Res. **242**, 606–619 (2015)
24. Zeng, X., Kang, S., Li, F., Zhang, L., Guo, P.: Fuzzy multi-objective linear programming applying to crop area planning. Agric. Water Manag. **98**, 134–142 (2010)

Towards the PhD Degree:
A Reflection and Discussion of the PhD Process

Marta Castilho Gomes[(✉)]

CERIS, Instituto Superior Técnico, Universidade de Lisboa,
Av. Rovisco Pais, Lisbon, Portugal
marta.gomes@tecnico.ulisboa.pt

Abstract. In recent decades the literature aimed at supporting postgraduate students in their PhD studies developed notoriously. This paper presents a reflection on managing the PhD process and accomplishing the doctoral thesis, based on this literature and the author's PhD experience.

The contents of presentations on this topic delivered to science and engineering audiences are first summarized (the latest having taken place at ICORES 2015 and 2016). Feedback from audiences is then described and the most significant issues raised in discussions listed. The concluding remarks reflect on the importance of disseminating this subject among the science and engineering academic community.

Keywords: Postgraduate students · PhD thesis and process · Supporting literature · Science and engineering

1 Introduction

Although each PhD is a unique piece of research work, there is much in common in the PhD experience independently of the area of study. In the decade of 1990 there was an explosion of literature directed at helping postgraduate students deal with their PhD process. This literature influenced me greatly during my PhD studies, namely the book by Phillips and Pugh, "How to get a PhD: A handbook for students and their supervisors". In recent years I have presented a reflection on the PhD process based on it, complemented by my personal experience, to science and engineering audiences.

The aim of these presentations is to motivate students to get a perspective of the PhD process at a higher level than their particular work, using the supporting literature to this end. By fully understanding the PhD process, regarding both the phases of research and the psychological stages they will go through, students can manage it in a more professional way. They will be able to cope better with difficulties, and so benefit from a smoother progress towards obtaining the degree.

This paper presents the reflection undertaken in the "PhD seminar" and accounts for the reactions received, being structured as follows. Section 2 looks at the development of the literature aimed at supporting PhD students, with particular emphasis on Phillips and Pugh's work, and describes how the seminar evolved. Section 3 deals with most of the seminar contents, while Sect. 4 addresses the specific topic of thesis writing. Section 5 presents the debate with audiences and other feedback I received. The last

© Springer International Publishing AG 2017
B. Vitoriano and G.H. Parlier (Eds.): ICORES 2016, CCIS 695, pp. 132–141, 2017.
DOI: 10.1007/978-3-319-53982-9_8

section is a final reflection on disseminating these topics among students and supervisors towards a successful PhD experience in science and engineering.

2   The Literature Supporting the PhD Process and the PhD Seminar

Although there were previously many good texts on the topic of academic research, the first books aimed specifically at PhD students appeared in the decade of 1990 [3, p.21]. Titles like "How to survive your doctorate: What others don't tell you" [6], "The unwritten rules of PhD research" [9] and "The smart way to your PhD: 200 secrets from 100 graduates" [1] hint that there are "secrets" the postgraduate student needs to know to succeed in his endeavour. In fact, a quick search online shows there are dozens of books available directed at unveiling such secrets to postgraduate students.

I was lucky to have been introduced to Phillips and Pugh's book "How to get a PhD: A handbook for students and their supervisors" by more advanced PhD students at my research centre. I bought the 2nd edition (1994), read it eagerly and tried to put the advice into practice: it had a profound effect on my PhD process. Grix says this is perhaps the best known textbook on PhD research [3, p.22]. The 1^{st}edition dates back to 1987 and the most recent one was published in 2010 (5^{th}edition), having been translated into several languages, including Portuguese.

I find there are three strong points in this book. One is its broad scope, not focusing in a particular area of studies. For instance, "Demystifying postgraduate research" [3] is also an excellent reference, however it is a book written for social sciences students. The second notable feature is the many real examples of students and supervisors, and the problems they faced, that the book describes. Real stories are always powerful to communicate a message and the book contains dozens of illustrative examples of students in Architecture, Engineering, History, Biology, Physics and other fields. The third aspect is the importance given to psychological aspects, which the real stories come to underline.

Influenced by Phillips and Pugh's book, as well as other references, during my PhD studies I decided to deliver a seminar on the topic of "How to get a PhD" for the postgraduate student community of IST (my school). Interest and feedback were very rewarding at the time, however I did not repeat the seminar until recently, having had the opportunity to present it in Portugal and abroad a few times in the past years. The timeline of the seminar is as follows:

2004: IST, Lisboa, as a PhD student (postgraduate students' community)
2012: Carnegie Mellon University, Pittsburgh(Chemical Engineering and Electrical and Computer Engineering departments)
2012: FEUP, Porto (Electrical and Computer Engineering department)
2013: FEUP, Porto (Electrical and Computer Engineering department)
2014: LNEC, Lisboa (National Laboratory of Civil Engineering)
2015: ICORES conference, Lisboa
2016: ICORES conference, Rome

Above, IST stands for Instituto Superior Técnico, Lisbon University, and FEUP denotes the Faculty of Engineering of Porto University (Portugal).

The next section presents the central topics of the seminar, which deal with the definition of a PhD thesis and the process a postgraduate student goes through to accomplish it.

3   The PhD Process and the PhD Thesis

Completing the PhD level of studies ensures that the student has attained the highest degree in a given field by successfully defending a piece of original research work that contributes to expand the frontiers of knowledge in that area. The ideas presented in this section are central points discussed by Phillips and Pugh [7] that greatly influenced me to achieve this goal and that I selected for the seminar from the vast material in the book. After describing them, much as they appear in these authors' work, I will give a personal view and interpretation, in light of my PhD experience ("my story"). The first three aspects mentioned are the most important ones that a postgraduate student should be aware of, as they might play a major role on how his PhD studies develop.

3.1   Most Relevant Topics (Based on Phillips and Pugh [7])

Phillips and Pugh discuss first in their book the nature of postgraduate education and the psychology of being a postgraduate. In fact, the candidate enters the system with the habits of an undergraduate student and leaves it as a doctor, which is a new identity. So, this is a process where the postgraduate student needs to adjust to a new reality.

In undergraduate education, teachers develop the contents for their courses and choose the bibliographical references, organize classes and laboratory work and give the final exams. The student is expected to fulfill the activity program previously established: in a word, to act in a *reactive mode*. In postgraduate education, it is up to the student to manage his learning and develop his research plan, the role of the supervisor(s) and other academics being to help the student achieve this aim. The student has thus the responsibility of deciding what is required and should not expect others to say what is the next thing to be done. He should discuss with the supervisor(s) what to do, what to read, what courses to take: to act with personal academic initiative in a *proactive mode*. This is indeed a very different style of operation compared to what the student experienced in undergraduate studies.

An intelligent student who completed his undergraduate studies successfully starts the new phase determined to make a relevant contribution to his field of studies. Focusing on the same subject for a large period of time, while performing many repetitive tasks and mingling in the postgraduate environment, will cause the initial enthusiasm to fade away. This is the natural evolution in the PhD process: in the end, the student emerges with the identity of a *professional researcher*. Phillips and Pugh characterize a researcher as being confident of his own knowledge but aware of its boundaries, someone who is able to argue his point of view with anybody regardless of status and, when needed, express a lack of understanding with confidence, and, finally, someone able to assess scientific work. So, with the PhD degree the candidate is recognized by a given scientific community to be an independent or autonomous researcher that can supervise others in the future.

The next vital issue is the concept of thesis. Nowadays the word is used to designate the final report regarding the PhD research project, but one should link the word to its Greek origin "thésis", meaning "to place": to place a view, an opinion that you wish to argue. While the concept of thesis as a position to be defended is certainly more familiar to students of social sciences and the humanities, I think it can be extremely useful for science and engineering students, too. In fact, as Phillips and Pugh [7] argue, a PhD thesis must have a clear "story line", with all parts adding up to a coherent argument, either in the way new data is presented and analysed or in the way existing data is interpreted.

The third and very meaningful aspect presented by the authors is the description of the psychological stages a PhD student will go through. Enthusiasm at first, then isolation and increasing interest in work, up to a point where student dependence is transferred from the supervisor to the work itself. At this stage, the need for external approval lessens and the student increasingly relies on his own judgement of the quality of his work. At later stages there will be boredom and even frustration, due to repetitive tasks and the inability to do everything the student would like to do in the available time. In the final stage, all the student can think of is finishing the thesis. Having finally submitted it, there will be anxiety while waiting and preparing for the final exam. And at last, when succeeding, the student will go through an exhilarating phase, with feelings of joy and achievement and a great increase of confidence. Phillips and Pugh [7] stress that only by being diligent, systematic and fiercely determined will a student complete his PhD research, and those qualities are far more needed than brilliance.

Recognizing and integrating these three aspects is, in my opinion, the cornerstone of a successful PhD experience. So, I greatly emphasize them in the seminar: to know how a postgraduate student should act and what is expected of him; to know what are the distinguishing features of a doctoral thesis; and to be aware of the psychological evolution in time during the PhD.

This is then complemented by some other aspects pointed out by Phillips and Pugh [7]. First of all, the notion that a PhD thesis must bring a contribution and that this should be rather limited in scope: "normal science", as opposed to paradigm or theory shift, which are not expected to be pursued during PhD research. Disruptive contributions are exciting but occur seldom; in the interim, ordinary research takes place and should encompass the contribution of a PhD thesis.

I use the examples of Albert Einstein and Karl Marx, given by Phillips and Pugh [7], because these two personalities are famous with any academic audience and therefore very significant to explain this point. Although they eventually gave rise to paradigm changes in their fields, Einstein and Marx first proved their competence as researchers by pursuing non-disruptive topics for their PhD theses that were very different from their later contributions: Brownian motion theory in the first case, the theories of two Greek philosophers (not the most widely known) in the second one.

The fact that the PhD contribution should be original is another problem, which worries students. This is explored in detail by Phillips and Pugh [7]: acknowledging that a discussion between students and supervisors on the issue of originality may be lacking, they list fifteen forms of being original. Without describing the list, I raise this key point so that the students can feel more reassured (being original may be easier than they thought at first) and motivated to go through the list later and be better prepared to implement original ideas in their own study.

Phillips and Pugh's work is very rich and in the seminar I can only make reference to a few points. Although it is of paramount importance, I do not have the time to go into the details of the chapter "How to do research", where the craft of doing academic research is discussed. However, I do address the stages of research and time management issues that the authors present. Typical stages in a PhD research are:

- Addressing the field of interest and enumerating possible topics, which culminates with the thesis proposal (this involves background theory and focal theory).
- Performing a pilot study, then full-scale data collection and data analysis (involving data theory).
- Although writing should be on-going since the first day, the third stage involves the final writing-up and fully developing the thesis contribution.

The student-supervisor relationship is also analysed by Phillips and Pugh. The authors go on to describe, on the one hand, what supervisors expect of their doctoral students, and on the other hand what students expect of their supervisors. Students should strive to inform their supervisors honestly on the evolution of the work, but also in a stimulating and creative way. By working on this relationship and fulfilling the supervisors' needs, the students can also be more successful in having their own needs and requirements satisfied.

Finally, finding a peer support group is essential so that the student can share ideas, emotions, the ups and downs of the process… until the final line is crossed!

3.2 My PhD Experience

Having presented these core aspects, in the seminar I explain how Phillips and Pugh's book, complemented by other material I read, had an influence on my PhD thesis and process. This can be divided into influence on the scientific work, on writing habits and on attitude. Most PhD students in science and engineering are perhaps unaware of the strong interplay between the first two aspects.

Doing a PhD in Systems Engineering ([2], "Reactive scheduling in make-to-order production systems: An optimization based approach"), I understood I did not have "research questions" to be presented and answered, linked to hypothesis that had to be tested, but rather I should present the "research gaps" in my field that justified the study. This was clearly stated in the thesis introduction, helping to properly set the objectives, and was discussed in detail in the final conclusion chapter. The notion of "storyline" helped me link the chapters (seven in total) in a smooth and logical way, so the thesis could be perceived as a coherent report on the work done.

Regarding the scientific work developed, I realized there are two crucial aspects: the postgraduate student must know both the standards in his field of studies and the standards of his university for a PhD thesis. Of course, it is essential to read relevant scientific papers (as many as possible), but the appointed aspects show that this is not enough. Early on in the process, students should have contact with several (many) PhD theses that are meaningful for their studies, particularly theses from the university where the student is enrolled. Careful analysis of these works will set a standard to be reached and become an inspiration for the student along the process.

The notion of thesis as a coherent study and the stages of the process defined by Phillips and Pugh [7] had implications in the design of my study. The first experimental chapter of my thesis was developed very much as a "pilot study": given the success of the optimization approach in this preliminary step, I then broadened its scope in subsequent research stages. This resulted in three other experimental chapters, where the last one compares the two types of job shop scheduling models developed (which differ in the way the time dimension is modeled).

The second aspect mentioned (how I managed writing) will be presented in detail in the next section. Finally, the supporting literature was also of great help in terms of attitude: it increased my confidence and sense of independence as a postgraduate student, which made discussions with my two advisors more fruitful for the work progress. It taught me to appreciate the value of my research environment and peer support group, leading me to contribute more actively to this group and also to benefit more from it.

4 Writing the Thesis

Efficient writing is critical for a successful PhD process, also in science and engineering. The seminar briefly addressed this topic, emphasizing the close link between written language and thought, which is one of the reasons for the difficulty experienced in writing [7]. After my ICORES 2015 presentation I was challenged to develop the issue of "writing the thesis" for ICORES 2016, and so explored the advice on how to become a proficient writer in the supporting literature. This section expects to motivate postgraduate students to this essential topic.

In Phillips and Pugh [7] the theme of writing the thesis appears in the chapter "The form of a PhD thesis" but, interestingly, in the 5th edition [8] there is an independent chapter after that one, entitled "Writing your PhD". The distinction is made between "serialists" and "holists": the former think and write sentences almost in final form while the latter think as they write and evolve by means of successive drafts. Advice on how to organize writing is given, where the baseline is that students have a writing schedule of two to five hours a week and stick to it, with no interruptions. This is especially important for students in science disciplines, many of whom prefer laboratory work and tend to procrastinate writing. The authors also emphasize the need to clearly formulate ideas in writing, which are familiar to the student but new to the reader, and give some rules for it. Notions of the appropriate style of writing should be acquired by reading academic journals and books in the field of studies.

Silvia [10] is an enjoyable book, rich in advice and captivating as promised by the title "How to write a lot: A practical guide to productive academic writing". The author discusses and deconstructs psychological barriers that impair writing, differentiating "binge" and "disciplined" writers. In line with the above authors, Silvia defends that it is by keeping to a writing schedule that inspiration eventually comes and "writer's block" is overcome: disciplined writing has many advantages and is the secret behind productive academic writing. Different sorts of motivational tools are proposed to the academic writer (postgraduate students included), which can take the form of simple statistical tools (histograms) that compare the number of words written per day to monitor progress.

Both Silvia [10] and Phillips and Pugh [8] mention the inspiring example of Anthony Trollope, a Victorian novelist who managed to write 63 books (some with more than one volume) while working full time at the post office. How did he do it? By writing everyday from 5:30 to breakfast time!

I found particularly interesting to learn in Silvia's book that much research has been done on how to write, and also that several famous writers have written about the subject. A science and engineering student may not have the time or will to go through this literature (Silvia is a professor of Psychology...), but should acknowledge that the process of writing is complex and worth studying in itself.

Grix [3] also offers valuable advice on writing a thesis. While acknowledging that writing strategies depend on personal preferences and previous experiences, the author warns postgraduate students against two mistakes: separating the writing-up phase from the research process in general, and hanging on to work for too long. As with Phillips and Pugh [7], in this author's opinion writing is a continuous process that the student ought to start as soon as he enrolls in a PhD program.

After presenting the former issues, I introduced a personal note in the ICORES 2016 seminar. Writing the thesis was definitely the hardest part of the work, compared to developing optimization models for job shop scheduling. I estimate that more than 50% of the time during my PhD was spent writing, which comprised three different aspects: (i) situating my work within the scientific literature (ii) presenting the work (iii) discussing the results and drawing conclusions.

While for my MSc I first completed the work and then wrote the thesis from scratch (and this worked fine), I knew I could not follow this approach in my PhD. So, during my PhD research, I was always writing: taking notes from what I read, planning meetings with my supervisors (for which I printed an agenda that handed them at start), preparing short progress reports and, of course, writing the thesis. After the first oral presentation of my work in a conference, I submitted it to an international journal, and this (with adaptations) became the first experimental chapter of my thesis. Fortunately, publication was fast and very encouraging. As the work developed, I wrote the other three experimental chapters, which I submitted either as journal papers or conference papers.

I found that writing continuously, in parallel with the work on optimization models, was valuable for the development of ideas and to attain a higher-quality report (thesis). The writing-up stage was longer and harder than I expected, and involved writing the chapters of the introduction and conclusion and the final version of the literature review, so as to improve the framework of the work and clearly establish its contribution.

To conclude, the "problem of writing the thesis" for PhD students in science and engineering is, in my opinion, not different from the problem of writing any other document! By gaining writing habits, students will become better at presenting their work in a way that is both effective and pleasant to read.

5 Reaction and Feedback from Audiences

Audiences listened to the seminar attentively and engaged in stimulating discussions, sometimes with fruitful follow-ups in individual conversations. This section summarizes the main aspects highlighted in these discussions. My general impression was that audiences in science and engineering were not very familiar with the topic of the supporting literature, but it made sense to them. Feedback from professors was richer because of their experience both with successful and unsuccessful PhD students.

In the debates after the seminar, the following aspects were raised:

- Doing a PhD research is a process of dealing with uncertainty and this was focused in the seminar, following Phillips and Pugh [7] when presenting the stages of the process. There is enormous uncertainty regarding the thesis content at start and no uncertainty when the final report is submitted! A professor underlined this aspect, which is inevitable, and said students should know and prepare to cope with this situation, which is stressful in itself.
- A professor related to the psychological stages of the process and said that during his PhD he felt actually depressed. A second aspect was that meetings with his supervisor were scheduled in late afternoon, after a day's work. If he was not enthusiastic about his research and made the meetings interesting, the supervisor would fall asleep!
- A student questioned self-help books and the presented approach as being useful, since doing a PhD is about mastering the research process in a particular area of studies, and this varies a lot between areas. I totally agreed with him regarding the nature of PhD research, but added that in my view being aware of general principles and common aspects of doing a PhD can be very helpful either as a postgraduate student or later as a supervisor. Of course, this is no substitute for knowing how research is done in a given field, either by taking courses and/or reading the specialized literature.
- The importance of publishing journal papers during the PhD research featured prominently in one of the discussions. This will be addressed in detail later.
- A professor had recently had a student who was very disappointed when she realized she probably would not find a job where the exact knowledge and techniques she developed in her thesis would be applied. So, he wanted to discuss how to motivate postgraduate students given that many, or most of them, will undergo similar experiences.
- The problem of changing supervisors also arose. This is a tricky academic issue, and in the end the opinion of the professor who introduced the matter was that students wanting to change supervisors should better move to another university.
- In one of the seminars I emphasized in the discussion that the PhD process is a human process, and so communication plays a vital role: to be successful the student should remember that at all times and be proactive and proficient in communicating with his supervisor(s), as well as with the academic community.

Later comments, either in face-to-face conversations or by e-mail, included the issues below:

- A professor acknowledged that "knowing that nobody will tell him what to do next" was the single most important thing the postgraduate student should be aware of.
- A student attending my first seminar at IST saw it as very positive that, as a postgraduate student, I decided to disseminate what I learnt from reading Phillips and Pugh's book to the student community.
- A professor noted that, in fact, most engineering students develop their PhD theses with no concern about the matters discussed in the seminar.
- A student who read Phillips and Pugh [8] at the start of his studies found it especially valuable to be conscious of what can go wrong in a PhD process, from the many concrete students' stories presented in the book.
- A student who had received the book "Student to scholar" [5] from his funding institution took it from the shelf and lent it to me. I read this relatively short book with great interest, and when giving it back to the student heard him say that, after listening to my seminar, he would pay attention to the book.
- A professor acknowledged the importance of the topic "writing the thesis".
- A pot-doc researcher said this sort of discussion is needed to prevent students from giving up their PhD studies (she knew of several cases, in her country).
- Finally, two professors told me, several months after the seminar, that it had an effect on the postgraduate student community. Students recognized the importance of discussing these issues and the presentation made them look at their PhD research in a different way, more mature, less naïve.

At FEUP in 2012 the discussion converged to the subject of publishing journal papers during the PhD research. In fact, starting in the decade of 2000, this has gained importance in Portugal for PhD studies, particularly in engineering. Validation of the work (or part of it) by the international community, as a complement to validation by the examining committee, is highly desirable. Publishing a paper in an international journal is a long process, while conference papers are usually easier to publish. PhD candidates should work towards the objective of having one paper accepted in an international journal when submitting the thesis. Of course, having more than one paper accepted in good international journals is even better, but difficult to achieve in the three to four year horizon of a PhD.

When concluding the seminar, and to further motivate students for these issues, I recommend the excellent short paper by Grover [4] "10 mistakes PhD students make in managing their program" and give out some paper copies.

6 Conclusion

By reading the currently available literature aimed at them, PhD students may understand the PhD process at a deeper level. As a result, they can structure their research and organize themselves better, gaining in effectiveness and self-confidence. Such literature is rich in information, useful advice and insights. It should be read at different stages of the PhD process, not only at the beginning, because students evolve and will benefit differently from it depending on where they stand. The book by Phillips and Pugh [7] influenced me greatly and became the basis for the seminar described in this paper. In my opinion, such a reflection on the PhD process is part of postgraduate education and, beyond the PhD stage, will enable students to guide others better in the future as supervisors.

An analysis of the literature supporting PhD students shows that these issues are far more discussed in management and social sciences than in science and engineering. Reaction of audiences to the PhD seminar confirmed that students and academics in science and engineering are still not familiar with the type of reflection and the "self-help approach" developed in such literature. However, there was evident interest and the discussions were lively. This strengthened my belief that it is important to disseminate this kind of literature and motivate the science and engineering world to benefit from the general principles and advice it conveys.

Acknowledgements. I am grateful to Professors Ignacio Grossmann, José Machado da Silva, Paulo Portugal and Joaquim Filipe, and to Pedro Marcelino, who invited me to deliver the PhD seminar to different audiences. They played an active role in its dissemination, while motivating me to keep on developing the ideas.

I also thank the many people who contributed with their comments to this reflection, namely Professor Dominique de Werra, who suggested I should look at the subject of "writing the thesis".

References

1. Farkas, D.: The Smart Way to Your Ph.D.: 200 Secrets from 100 Graduates. Your Ph.D. Consulting: USA (2009)
2. Gomes, C.M.C.P.S.: Reactive scheduling in make-to-order production systems: An optimization based approach. PhD thesis in Systems Engineering, Instituto Superior Técnico, Universidade Técnica de Lisboa (2007)
3. Grix, J.: Demystifying Postgraduate Research: From MA to PhD. University of Birmingham Press, Birmingham (2001)
4. Grover, V.: 10 mistakes PhD students make in managing their program. Decision Line, May, 11–13 (2001)
5. Levasseur, R.E.: Student to scholar: The guide for doctoral students. MindFire Press, Florida (2006)
6. Matthiesen, J., Binder, M.: How to survive your doctorate: What others don´t tell you. Open University Press, Berkshire (2009)
7. Phillips, E.M., Pugh, D.S.: How to get a PhD: A handbook for students and their supervisors, 2nd edn. Open University Press, Bukingham (1994)
8. Phillips, E.M., Pugh, D.S.: How to get a PhD: A handbook for students and their supervisors, 5th edn. Open University Press, Berkshire (2010)
9. Rugg, G., Petre, M.: The unwritten rules of Ph.D. research. Open University Press, Berkshire (2004)
10. Silvia, P.: How to write a lot: A practical guide to productive academic writing. American Psychological Association, Washington DC (2007)

Competition and Cooperation in Pickup and Multiple Delivery Problems

Philip Mourdjis[1]([⊠]), Fiona Polack[1], Peter Cowling[1], Yujie Chen[1], and Martin Robinson[2]

[1] YCCSA & Department of Computer Science, The University of York, York, UK
pjm515@york.ac.uk
[2] Transfaction Ltd., Cambridge, UK

Abstract. Logistics is a highly competitive industry; large hauliers use their size to benefit from economies of scale while small logistics companies are often well placed to service local clients. To obtain economies of scale, small hauliers may seek to cooperate by sharing loads. This paper investigates the potential for cost savings and problems associated with this idea. We study dynamic scheduling of shared loads for real-world truck haulage in the UK and model it as a dynamic pickup and multiple delivery problem (PMDP). In partnership with Transfaction Ltd., we propose realistic cost and revenue functions to investigate how companies of different sizes could cooperate to both reduce their operational costs and to increase profitability in a number of different scenarios.

1 Introduction

With over six thousand hauliers in the UK alone [15], competition is fierce. Hauliers face the orthogonal demands of short notice from customers, an expectation of low-cost service, and environmental sustainability concerns [12,21,24]. Because larger carriers can leverage economies of scale to benefit in routing and scheduling, competition is getting ever stronger. If smaller carriers could work together, they could increase scheduling efficiency, save on mileage costs, and improve flexibility. In this paper we quantify the savings possible when carriers outsource some of their customer consignments to other carriers, working either independently or as a group.

As a real-world problem, there are constraints that must be satisfied, such as vehicle capacity, soft time windows and driver working hour rules. The problem is defined in terms of consignments which include a single pickup location and one or more delivery locations. Consignments vary in size, and may be able to share one delivery vehicle, to save cost. A key constraint is that each vehicle must be unloaded in the reverse order to the loading order: deliveries from one vehicle are constrained to a last-in, first-out (LIFO) order. Concretely, consignment A may be interrupted by another if all of the second consignment's deliveries are serviced before continuing with consignment A's deliveries. We call this a *pickup and multiple delivery problem* (PMDP). This paper investigates the cost savings which are possible if carriers distributed across a country share consignments.

© Springer International Publishing AG 2017
B. Vitoriano and G.H. Parlier (Eds.): ICORES 2016, CCIS 695, pp. 142–160, 2017.
DOI: 10.1007/978-3-319-53982-9_9

2   Related Work

Research on PDPs usually concentrates on static models of small scale problems such as servicing taxi requests, or ride sharing schemes [29] – dial-a-ride problems (DARPs). [13] present a widely accepted mathematical formulation for the generic PDP, which they refer to as the *vehicle routing problem with pickup and delivery and time windows*.

Variations of the PDP handle constraints on the number of vehicles used, time windows on requests, capacities and number of depots. However, most of the existing research is on static problems, in which all requests are known in advance [4]. Exact solutions to static PDPs favour branch-and-cut-and-price algorithms using column generation techniques, for example, [16] uses this approach to solve a multi-depot PDP for problems with up to 55 requests. No indication is given of whether their approach scales to larger problem sizes.

Exact solutions to dynamic problems include a variation of the column generation approach [19], used to solve DARPs of up to 96 requests, with either static or dynamic time windows. [30] solve a PDP based on real-world logistics with multiple carriers, vehicle types and LIFO constraints using a set partitioning formulation containing an exponential number of columns. However, in general, exact methods do not scale well, so heuristic and hyper-heuristic approaches that can quickly find near-optimal solutions, have become popular for large-scale, real-world problems. [20] provides a good overview of exact and heuristic methods for vehicle routing problems. More recently, heuristic approaches have been applied to scheduling with LIFO loading constraints [3,9,11].

[9] use a three phase approach. First, multiple routes are created using a greedy randomised adaptive search procedure; next variable neighbourhood descent (VND) applies local search to derive new solutions using a diversification strategy derived from [26]. Finally, crossover is used to combine solutions to form further candidate solutions. [3] use a multi-start tabu search approach that uses Clarke and Wright savings [10] as well as two random schedule heuristics to build seed routes. The tabu search improves solutions by repeatedly removing and re-inserting consignments, using traditional strategies to prevent cycling and promote diversification.

Existing approaches to dynamic scheduling of PDPs (summarised in [8]) often use a two-phase hyper-heuristic [5]: requests are first inserted into a schedule, then optimisation is performed, either on a route that has been changed or on an entire schedule. Research has focused on different insertion, removal and local search operators, and on the heuristics that choose between operators at any point. For example, [17] use neighbourhood search heuristics and ejection chains to tackle same-day courier PDP. [22] use a double horizon approach with routing and scheduling sub-problems to schedule similar problems of a larger size. [1] use probabilistic information to inform their routing of a multi-period VRP.

We are concerned with efficient solution of scheduling under just-in-time logistics, where the customer expectation is that hauliers respond quickly to delivery requests, and where same-day delivery often attracts premium payment rates. In the traditional approach used by small haulage companies, static scheduling is re-run daily. However, static scheduling cannot be used for real-time response to orders, and does not take account of the existing sched- ule and loading. We propose a dynamic scheduler that intelligently adapts to incoming requests, a novel variant of dynamic PDP [5].

Our model of the PMDP is based on the generic PDP model of [13]. Our variable neighbourhood descent with memory (VNDM) hyper-heuristic takes inspiration from the hybrid variable neighbourhood tabu search (VNTS, [2]), which outperforms tabu and variable neighbourhood approaches for static VRPs. A schedule is built up by repeatedly inserting requests then performing optimisation. The strict LIFO constraint in PMDP, along with constraints such as the vehicle capacity, makes it difficult to find improving moves in PMDP, so we develop a descent based algorithm and local search operators tailored to PMDP, with roots in classic VRP and PDP solutions. Once a solution has been built, we perform optimisation whilst aiming to minimise ordering inversions within a vehicle's schedule, as these are unlikely to improve results in problems with tight time windows and LIFO constraints on deliveries. Local search techniques that affect delivery order, such as those presented by [28] and [7], and the GENI technique [18], are unsuitable for direct use on our problem because they cause large changes in schedule ordering.

3   Model

The PMDP is defined on a directed graph $DG = (N, A)$ where A is the arc set and N is the node set. Each request r is identified by $(n_r, l_r, [t_r^{start}, t_r^{end}], tt_r^{service})$ where n_r is the location, l_r is the load (where the summation of pickup load and delivery loads for a consignment is equal to zero). $[t_r^{start}, t_r^{end}]$ represents the start and end times of the arrival window respectively where the service time $tt_r^{service}$ must begin (for clarity we use double letters to represent quantities). R is the set of requests where $R = P \cup D \cup O$, P being the set of pickup-requests and D the set of delivery-requests. O is the set of origins which are dummy requests used to represent the multiple depot locations of the problem. The arc between two requests r and u (that is, between nodes (n_r, n_u)) is the arc (r, u). A consignment c is identified by (p_c, D_c, t_c) where p_c is the pickup-request and $D_c = d_c^1, \ldots, d_c^{nc}$ is the sequence of delivery-requests. Each consignment has a received time t_c, which is the time at which the order is entered in the system. C is the set of consignments. $A_k \subset A$ represents the feasible arcs for vehicle k. The binary flow variable b_{ruk} is set to one if arc $(r, u) \in A_k$ is used by the vehicle k, and to zero otherwise. ll_{rk} is the load of vehicle k at request r and is not fixed but dependent on the other arcs in the vehicle's route. It is calculated as a running sum where each request either adds to the load (pickup) or subtracts from the load (delivery). A vehicle starts and ends its route at one of the depots with load equal to zero.

The goal is to minimise the total cost of servicing all requests $r \in R$:

$$\min \sum_{k \in K} \sum_{(r,u) \in A_k} C_{ruk} * b_{ruk} \tag{1}$$

where:
$$C_{ruk} = nc(nn_{ru}, ll_{rk}) + tc(ruk) + dc(tt_{uk}^{delay}) \tag{2}$$

subject to the constraints in Sect. 3.1. C_{ruk} is the cost of vehicle k servicing (r, u), calculated using running cost estimations for a 44-tonne articulated truck based on 2014 data from the UK Road Haulage Association (RHA) [14]. The component costs are: $nc(nn_{ru}, ll_{rk})$, the cost of travelling distance nn_{ru} (the length of arc (r, u)) with load ll_{rk}; $tc(ruk)$, the cost of the time taken by vehicle k to travel arc (r, u); and $dc(tt_{rk}^{delay})$, the cost of the penalty for arriving late at request u. We use a stepwise function (increasing every hour) after an initial grace period, in line with industry practice. Consignments may be either customer orders or backhauls (post-delivery return to pickup location, for instance to dispose of packaging), these differ only in that backhauls are usually mostly empty loads.

3.1 Constraints

The constraints for the PMDP are laid out in Tables 1 and 2. The constraints in Table 1 have been adapted and expanded from the formulation for the PDP presented by [13]; Table 2 presents the additional new constraints for the PMDP.

Table 1. Adapted constraints from [13], here $\equiv$ implies that this constraint is equivalent to a constraint presented by Desaulniers et al. and $*$ implies that this constraint has been modified for the PMDP.

$\equiv$	$\sum_{k \in K} \sum_{u \in R_k} b_{ruk} = 1 \ \forall r \in R$	(3)
$*$	$\sum_{u \in P_k} b_{ruk} * |D_j| - \sum_{w \in D_j} b_{rwk} = 0 \ \forall k \in K, r \in R_k$	(4)
$*$	Removed	(5)
$*$	Removed	(6)
$*$	Removed	(7)
$\equiv$	$b_{ruk} \left(t_{rk} + tt_r^{service} + tt_{ru} - t_{uk} \right) \leq 0 \ \forall k \in K, (r, u) \in A_k$	(8)
$*$	$t_r^{start} \leq t_r^{end}, t_r^{start} \leq t_{rk} \ \forall k \in K, r \in R_k$	(9)
$*$	$t_{rk} + tt_r^{service} + tt_{ru} \leq r_{uk} \ \forall k \in K, r \in P_k, u \in D_r$	(10)
$\equiv$	$b_{ruk} \left(ll_{rk} + l_u - ll_{uk} \right) = 0 \ \forall k \in K, (r, u) \in A_k$	(11)
$*$	$0 < l_r \leq ll_{rk} \leq l_k \ \forall k \in K, r \in P_k$	(12)
$*$	$l_r + \sum_{u \in D_r} l_u = 0 \ \forall r \in P$	(13)
$\equiv$	$l_o(k) = 0 \ \forall k \in K$	(14)
$\equiv$	$b_{ruk} \geq 0 \ \forall k \in K, (r, u) \in A_k$	(15)
$\equiv$	$b_{ruk} \text{ binary } \forall k \in K, (r, u) \in A_k$	(16)

Constraints (3) and (4) ensure that each arc is only included once and that a pickup and all its corresponding deliveries are handled by the same truck. Here, $|D_u|$ is the number of delivery-requests for pickup-request u. Constraint (4) is non-standard for the PDP and is necessary as there may be multiple delivery-requests per pickup-request. It states that for each pickup request there exists a

Table 2. New constraints for the PMDP.

$$|P_c| = 1 \; \forall i \in I \tag{17}$$
$$|D_c| \geq 1 \; \forall i \in I \tag{18}$$
$$t_{rk} < t_{uk} \; \forall k \in K, r \in P_k, u \in D_r \tag{19}$$
$$t_{rk} < t_{uk} \Rightarrow t_{vk} < t_{wk} \; \forall k \in K, \; \forall r, u \in P_k, \; \forall v \in D_u, \; \forall w \in D_r \tag{20}$$
$$\sum_{(r,u) \in A_k} b_{ruk} \left(tt_r^{service} + tt_{ru} \right) \leq tt_k \; \forall k \in K \tag{21}$$

$b_{ruk} = 1$ and that this multiplied by the number of deliveries is the same as the number of arcs that end at each of the corresponding delivery requests. Unlike [13], we are not interested in multicommodity flow, so we omit constraints (5) to (7). Constraint (8), imposing total schedule duration, remains unchanged. Constraints (9) and (10) have been modified to allow for soft time windows. Constraints (11) to (13) specify that a pickup node must have positive load and that deliveries must have negative load, also that the sum of pickup and delivery loads is zero. The initial vehicle load, non-negativity and binary requirements are the same as [13]. The following constraints have been added for the PMDP: (17) and (18) specify that a request has exactly one pickup and may have arbitrarily many deliveries. (19) specifies the precedence between a pickup and its deliveries while (20) expresses the LIFO constraint. Finally, (21) specifies that the length (in time) of any vehicles route is less than a value E_k which may be set according to local conditions.

Minimising k, the number of vehicles used, is not considered as part of this problem, though it is kept low as a side effect of the heuristics used. For each truck, requests may be nested within other requests if LIFO and capacity constraints are not violated.

4   Solution Approach

Our PMDP solution, Like other hyper-heuristic approaches, is a two-phase process. An initial set of routes is built using a greedy constructive heuristic and then optimised with the variable neighbourhood descent with memory (VNDM) hyper-heuristic. This manages a set of low level heuristics (LLHs), introduced in Sect. 4.3.

4.1   Constructive Heuristic

As consignments enter the system dynamically and are not known in advance, the insertion heuristic treats each consignment atomically, finding the lowest cost insertion location across all routes for a pickup and all its deliveries (guaranteeing LIFO), such that no previously inserted consignment incurs a delay. This process is a greedy exhaustive search over all potential insertion locations and the position with the lowest cost is chosen.

4.2   VNDM Hyper-heuristic

After the insertion of each new consignment, VNDM is used for optimisation, running for a constant amount of CPU time. A fixed CPU time is used as there is no need to find a global optimum when new consignments that arrive will force changes to any schedule created. VNDM is a descent-based first-improvement heuristic. Routes are first ordered by length, then each LLH generates a list of potential moves. Since the majority of a schedule is unaltered after a modification, VNDM limits revisiting parts of the search space by maintaining a record of LLHs that give no improvement on each route (pairs of route and LLH identifiers are stored). If a LLH fails to produce an improving move, it is added to a tabu list. The tabu list is re-initialised when a route is subsequently modified, as a LLH may now be able to find improvement where none was previously possible.

VNDM differs from other published PDP solution approaches in a number of ways, notably in the choice of local moves used (specific to the PMDP), the use of route ordering to focus the search on promising areas, and the use of a route memory to reduce repeated searching. The search space is further reduced by imposing distance and time limits on nodes chosen for potential moves, which are different for each LLH and determined through extensive testing.

4.3   Low-Level Heuristics

The nature of PMDP, with strict LIFO ordering of consignments, guides our selection of LLHs to apply to route optimisation. Since a pickup request must occur before its delivery requests, reversing a section of a schedule and repairing infeasible pickup / delivery ordering will significantly alter the distance of the route. Because time windows are usually tight, increased distance may result in significant delay in servicing requests.

In selecting LLHs to modify routes, a consignment may only be rescheduled if the modification results in a valid schedule. A consignment may be scheduled such that other pickups or deliveries occur between the consignment's pickup and final delivery, providing load and LIFO constraints are not violated. However, if the consignment is rescheduled, the nested pickups or deliveries from other consignments remain in the original schedule, thus allowing modifications to undo nested consignments.

Highly disruptive LLHs that introduce partial route inversions cannot improve our schedules as these would invalidate either the LIFO or precedence constraints of pickups and their deliveries. This rules out LLHs such as GENI [18] and iCROSS [6]. However, we can use the CROSS exchange of [27] (used by [28]) as it does not reverse chains of requests. Additional LLHs, such as GENI-PO [23], have been chosen or developed to preserve existing schedule ordering as much as possible. By keeping the pickup and deliveries of one consignment in the same schedule (rather than splitting the consignment across loads and using precedence constraints), we facilitate the use of LLHs from the widely-researched area of one-many-one VRPs [7].

We provide four LLHs that can be applied to a single route. If a single route operator can generate more than one resulting route, that which is least disruptive to existing schedule ordering is used. LLHs that would reverse the order of a chain of requests are not allowed, hence we do not use 2-Opt.

3-Opt moves one consignment to a different position in the route schedule, whilst *4-Opt* swaps the positions of two consignments in the route schedule. *Nest Consignment* moves a whole consignment to a position within the delivery schedule of another consignment, thus nesting the first consignment within the second. Finally, *Nest Two Consignments* nests two consignments inside other consignments, a useful move where single-level nesting produces no improvement.

We provide four further LLHs that operate on more than one route at a time. GENI-PO [23] is a non-inverting variant of GENI [18]. The other three LLHs are from [27]. *Relocate* moves one consignment to a valid position in a different route schedule, which may introduce nesting. *Geni-PO* is a variation of relocate that preserves as much previous ordering as possible by moving a consignment to be geographically close to other consignments: all possible insertion position pairs are considered to find the most improving relocation. *Swap* exchanges consignments from two different routes, whilst *Cross* exchanges two chains of consignments between routes, preserving the existing ordering within each chain. Cross considers chains of all lengths when used.

Use of Local Moves. Of the eight LLHs, three consume only small amounts of CPU time for problems of the size we study (3-Opt, 4-Opt and Nest consignment), whilst the others (Nest two consignments, Relocate, Geni-PO, Swap and Cross) are considered *hard* and take a significant amount of time. However, the hard LLHs generate several orders of magnitude more potential moves than the computationally trivial moves. There is no intuitive reason to prefer one hard LLH to another, and there is little advantage to running more than one hard LLH at a time. Thus, to prevent VNDM optimisation simply running out of time whilst applying too many hard LLHs, each call of VNDM uses a neighbourhood structure comprising the three low-CPU LLHs in the order above, then one hard LLH, selected at random. The random selection ensures that all the hard LLHs are used over a series of optimisations, and thus provides ample diversification.

5   Computational Results

In collaboration with Transfaction Ltd., we have access to real scheduling data and manually-scheduled consignments for small UK hauliers (referred to as *real data*). The real data are insufficient, in quantity and quality, for our scheduling research, but provide us with indicative distributions and other information, from which we generate larger, realistic, data sets on requests and consignments (referred as *generated data*).

We generate 100 scenarios from a data set of 27,153 real-world consignments. The scenarios are built by selecting 200 real consignments at random from this set and building pairs of consignments representing outbound *linehaul* and return *backhaul* legs. Each consignment consists of at least two requests.

Initially, each haulage company (carrier) is assumed to have an unlimited number of vehicles and is represented by a depot, randomly located within the area encompassing the consignments. Consignments are assigned to the carrier, from the set of carriers with the fewest consignments, that is geographically closest to the midpoint between a consignment's pickup and final delivery locations. Thus, the initial schedule systematically distributes consignments evenly across many carriers. To analyse a dynamic system in silico, we use discrete event simulation.

5.1   Discrete Event Simulation (DES)

DES [25] is used to simulate the dynamic receipt of consignment requests. In order to add new consignments to a schedule that is already being serviced, we keep track of simulation time (an internal representation of current time, stored so that requests which in reality would have already happened cannot be modified by our optimisation procedure). If the scheduled start time of any request is before the current simulation time, it is marked as "fixed". Additional requests cannot be inserted before these fixed requests, and the routing of a fixed request cannot be altered in any optimising moves.

For each experiment we simulate one dynamic scheduling week, and limit optimisation to 5 min of CPU time. Each scenario is run 30 times, using a heterogeneous cluster of Intel Xeon based servers, totalling 72 cores and 120GB of RAM. The results presented here thus represent thousands of CPU hours.

5.2   Simple Cooperative Strategies

The first set of results compares the average per request costs for five carriers, exploring the effect on one carrier (the sample) under four different configurations of cooperation with the other four carriers. *All Contracted* has each consignment assigned to a specific carrier. Optimisation is only possible between vehicles belonging to the same carrier. *Out-sourcing* starts with a competitive model, but allows re-assignment of consignments from the sample to any of the other carriers, if cost savings can be made. *Out-sourcing to coop(erative)* adds the out-sourcing model for the sample carrier into a model in which the other carriers can exchange consignments if savings can be made; the sample carrier does not accept any additional consignments. Finally, the *cooperative* model initially assigns all consignments to individual carriers (as in *Contracted*) but allows unrestricted re-allocation during optimisation, if cost savings are possible.

The costs presented in Fig. 1 show that for the sample carrier, an average 9% saving can be made by out-sourcing to the four other carriers, whilst the configuration that allows other carriers to also cooperate results in average savings of nearly 14%, because the cooperation allows more efficient routing across the carriers. If the sample carrier also cooperates in efficient scheduling, the total

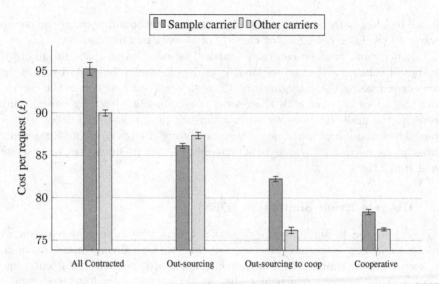

Fig. 1. Average cost for a single carrier (sample carrier) and a group of carriers (other carriers), with four different models of cooperation.

average saving for the sample carrier rises to 18%. Cooperation is also beneficial for the other carriers: accepting orders from the single carrier can produce benefits of 3%, whilst cross-group cooperation produces savings to averaging 15%.

The results shown should drive all carriers towards cooperation. Competition favours carriers with the lowest costs; the sample carrier achieves this in configuration 2, by outsourcing to other carriers who are not cooperating. However, rational competitors would be expected to copy this behaviour, moving the system towards a reallocation of consignments as seen in configuration 3; here, the competitors are cooperating, and the sample carrier is at a competitive disadvantage. However, if all carriers cooperate, as in configuration 4, the lowest costs for all carriers are observed.

Increasing cooperation allows a greater number of consignments to be handled. Figure 2 shows that the schedule in which all carriers operate alone covers on average less than 70% of their assigned consignments. However, the fully cooperative model can schedule over 85% of consignments. (Note that random scenario generation means that there is no guarantee that all consignments are feasible given the number of carriers, their locations and that even with an infinite number of vehicles, some consignments are too far apart to be serviced whilst adhering to driver working hour rules: since we do not consider driver sleeping arrangements and all routes must begin and end at the depot, these consignments are impossible in our current model.)

Table 3 shows the percentage of consignments that are re-allocated from the sample carrier in each configuration. Both out-sourcing and out-sourcing to a cooperative allow almost two-thirds of the carrier's consignments to be assigned

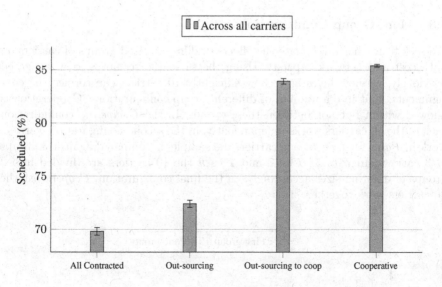

Fig. 2. Percentage of assigned consignments serviced across the four different models of cooperation.

Table 3. Percentage of the sample carrier's consignments re-allocated in different configurations.

Config	Cooperation Mode	Re-allocated
1	Competitive	0%
2	Out-sourcing	65.6%
3	Out-sourcing to coop	67.2%
4	Cooperative	57.2%

to others: because our scheduling algorithm minimises cost, these re-allocations can be interpreted as being carried more cheaply, due to more efficient use of resources, when assigned to other carriers. We are most interested in the percentage of consignments that are re-allocated away from the sample carrier. When outsourcing and cooperation are combined (configuration 3), the sample carrier's re-assigned loads are most cost-effective, as, in this configuration, the other carriers can also re-allocate loads among themselves (but not to the sample carrier). In the fully cooperative model, the sample carrier's consignments are less cost-effectively reassigned than in other reallocation configurations. However, the overall cost-effectiveness of the 5 carriers is significantly better than in other configurations: 62.5% of other carriers' consignments were reallocated in this model, leading to the reduction in cost observed for cooperation in Fig. 1. These results also strongly support the contention that savings can accrue to small hauliers who cooperate to carry each others' consignments efficiently.

5.3   More Group Configurations

We seek to further investigate the effects of different sized groups of carriers on both cost and network capacity. Using the same 100 scenarios as investigated previously, we now investigate how efficiently 10 carriers can service the consignments, split into a number of different group configurations. Cooperation is allowed within but not between these groups. In the *Competing* configuration each of the 10 carriers works independently, in the second configuration, carriers work in *Pairs*. In *1 vs 3s*, one carrier, the sample, is compared against 3 groups of 3 carriers. In *5 vs 5*, *3 vs 7* and *1 vs 9* the 10 carriers are divided into 2 groups of differing sizes accordingly. In the final configuration, *Cooperative*, the 10 carriers work together.

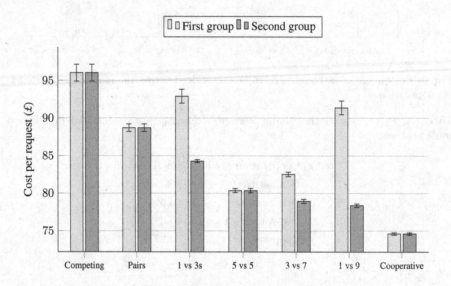

Fig. 3. Cost per request for different carrier group configurations.

Figure 3 confirms our earlier findings that working as a group can substantially reduce costs and additionally shows that larger groups can attain bigger cost reductions than smaller groups.

In each configuration, consignments are divided equally between groups, not carriers, such that, for example in the *1 vs 3s* configuration each group of carriers is assigned 100 consignments out of 400 but in the *1 vs 9* configuration, each group is assigned 200 consignments. Because of this, carrier 1 has more choice in the 1 vs 9 configuration and can achieve slightly better results than in the 1 vs 3s configuration, however the number of consignments that can actually be served is dramatically reduced as can be seen in Fig. 4.

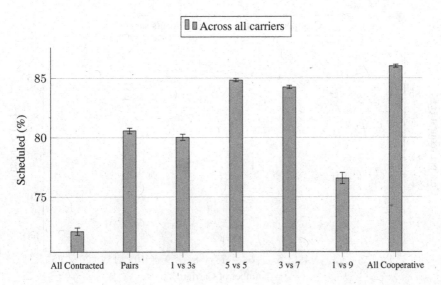

Fig. 4. Percentage of scheduled consignments for different carrier group configurations.

Figure 4 shows again the increase in network capacity made possible through cooperation. It is also clear that the largest savings are made quickly: just pairing with one other carrier can increase the number of scheduled deliveries from 72% to 80%.

5.4   Carrier Group Size

Extending our analysis, we seek to identify if there are diminishing returns for increasing the number of carriers in a cooperative group. Figure 5 shows how both the cost per request and the percentage of consignments scheduled improve as the size of a cooperative group increases. Though there are linear savings evident above 10 carriers, the majority of benefit is found between 1 and 5 carriers. These results must be qualified by stating that our consignments cover the UK and our carriers are randomly located across this area; since distance costs are a dominant factor in real-world pricing; if larger distances are involved, for instance across Europe, America or Asia, a larger number of well distributed carriers would likely be necessary to produce these savings. These results can be thought of more as suggesting that 10 major transport hubs is sufficient for efficient vehicle routes in the UK.

So far, we have assumed an infinite number of vehicles at each carrier location; in practice there will be a limited supply of vehicles at each carrier and therefore multiple carriers in the same area would need to work together. The following section investigates cooperation in resource constrained situations.

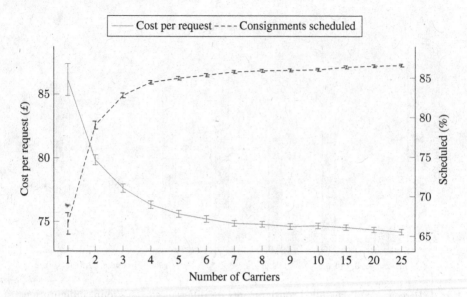

Fig. 5. Cost per request and number of consignments successfully scheduled as the number of carriers working together increases.

5.5   Competition

The experiments so far have assumed an infinite number of vehicles available for all carriers and looked at cooperation from the assumption that all companies work together to reduce total costs. We use the same 100 scenarios each with 10 carriers and 200 orders (assigned as before). However, companies now have a fixed number of vehicles. If a company cannot satisfy an order assigned to it, instead of creating additional vehicles, the customer is re-assigned to a random company that can service it. This means that a better utilisation of assets will lead to more customers for a given company. We also introduce a model for order revenue, enabling us to estimate carrier profitability. We assume that companies will not share their orders if it results in them loosing money, therefore, when cooperation is allowed between two companies, the company originally assigned an order always receives the profit it would make. For a different company to fulfil this order, it must yield sufficient profit to pay off the original company and still cover the associated delivery costs. The revenue model for an order is:

$$\text{Revenue}(c) = l_{p_c} \sum_{r \in D_c} nn_{r-1,r} \tag{22}$$

where the revenue of consignment c is a linear function of the total distance between all requests in the consignment multiplied by the pickup load. A company's total profit is the revenue of all the consignments it delivers minus the total cost of serving these, as specified in Sect. 3.

We now consider variants of the scenarios previously investigated, but, in each case, the number of vehicles is fixed at 40. We consider the impact of cooperation in scenarios with different distributions of these vehicle between the 10 companies, to simulate different competitive environments.

Equally Sized Companies. First, to validate our previous findings the 10 companies are set to have equal size, with 4 vehicles each. As expected, Fig. 6 shows that cooperation increases the profitability of all companies.

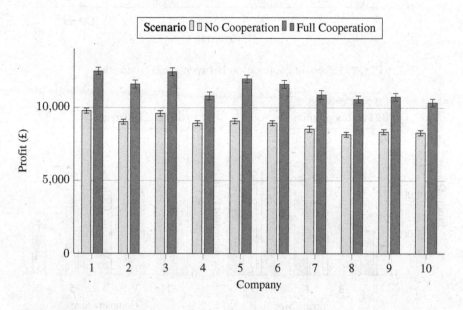

Fig. 6. Effect of cooperation on the total profits for ten equally sized companies.

Differently Sized Companies. The ten companies are now given different numbers of vehicles, set to: "2, 2, 3, 3, 4, 4, 5, 5, 6 and 6" respectively. Figure 7 shows the increased profitability of the first three companies when they work together as a cooperative assuming that all other companies continue to work independently. Profit increases of 12–18% demonstrate that even the smallest companies benefit from cooperation.

Looking at the group of heterogeneously sized companies in more detail, Fig. 8 shows how company size affects both raw profitability and the benefit of cooperation. Larger companies are able to produce more optimal routes and service more customers, generating more profit. When all parties cooperate, the profits for companies of all sizes increases. We can see that, as a percentage, small companies stand to gain the most from working cooperatively, with gains of up to 50%. Compared to the 12–18% result, above, it is again clear that more companies working together produces better results.

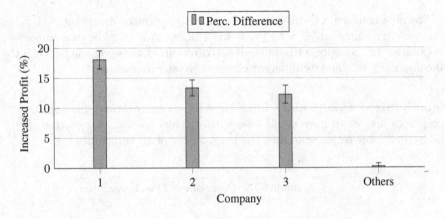

Fig. 7. Effects of cooperation between small companies.

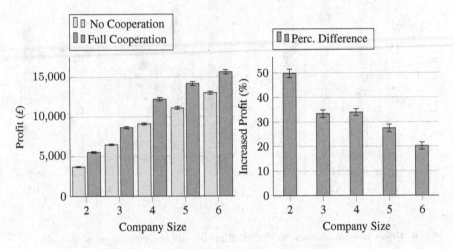

Fig. 8. Effects of cooperation across different sized companies.

Large Vs Small Companies. In this scenario we compare the profitability of large and small companies competing in the same market. The ten companies are now set to: "8, 2, 2, 2, 2, 2, 2, 2, 9, and 9" vehicles respectively. Figure 9 shows that, initially (when no companies are cooperating), the 2 largest companies produce the most profit. When the small companies work together they can increase their profitability and reduce the profits of the larger companies. Finally we observe that if the first large company joins the cooperative it can massively outperform its competitors. Other companies' profits fall, and the cooperative can more effectively handle orders (so orders are not stolen by the large companies).

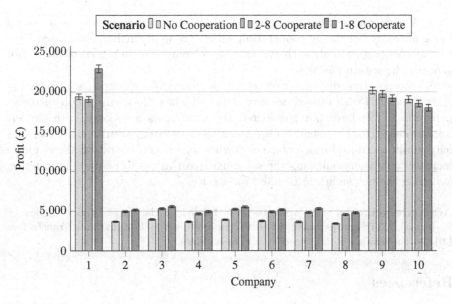

Fig. 9. Profit in a scenario with large and small companies. Company 1 has 8 vehicles. Companies 2–8 have 2 vehicles each. Companies 9 and 10 have 9 vehicles each.

6   Discussion and Conclusions

We have presented the VNDM hyper-heuristic as an effective schedule optimisation for PMDP under dynamic consignment requests. The objective and ordering constraints of the problem are set out and a set of LLHs optimised for these is given. We use data from the RHA to explore pricing and marginal costs of consignments, and show that cost savings of 15% to 18% are possible when hauliers cooperate. Cooperation also increases the capacity of a group of hauliers, by as much as 21%. The benefits of cooperation see diminishing returns above 10 separate carrier locations working together assuming sufficient numbers of vehicles to meet demands. Larger cooperatives will always have lower operating costs than smaller ones as they are able to more efficiently schedule their consignments to the most optimal company locations.

We have carried out further investigation into how savings from cooperation could be turned into increased profit in resource constrained problems with a fixed number of vehicles. We propose that the revenue of a customer be modelled as a linear combination of distance and load and define company profit as the sum of revenues over all delivered consignments minus the costs associated with delivering these loads. We consider that each company aims to maximise its own profit by only reassigning customers when a cooperating company can pay off the original company's profit and still cover its delivery costs. The cooperating company makes the cost saving as its profit on such orders. We have shown that this more realistic model of cooperation still leads to increased profits for all cooperating parties in a variety of different scenarios with differing company sizes. A particularly interesting result is that competing large companies stand to significantly benefit by cooperating with a group of smaller companies. Benefits of cooperation scale with the number of companies in the cooperative but generally lie within 15–20%.

We do not consider issues of vehicle reliability, for example, who pays the costs associated with missed delivery slots and what effect this has on customer perceptions. We have not considered the fixed costs associated with carrier-owned vehicles in this research; implementing the strategies outlined in this paper may result in reduced usage of carrier owned assets as cooperation allows for an increase in capacity, allowing the same fixed cost assets to be more productive, assuming there is sufficient demand for service.

Acknowledgements. This work has been funded by the Large Scale Complex IT Systems (LSCITS) project of the EPSRC. The authors would like to thank Transfaction Ltd. for the real-world data used in our experiments.

References

1. Albareda-Sambola, M., Fernández, E., Laporte, G.: The dynamic multiperiod vehicle routing problem with probabilistic information. Comput. Oper. Res. **48**(1), 31–39 (2014). http://linkinghub.elsevier.com/retrieve/pii/S0305054814000458, http://dx.doi.org/10.1016/j.cor.2014.02.010
2. Belhaiza, S., Hansen, P., Laporte, G.: A hybrid variable neighborhood tabusearch heuristic for the vehicle routing problem with multiple time windows. Comput. Oper. Res. **52**(part B), 269–281 (2013). http://linkinghub.elsevier.com/retrieve/pii/S0305054813002165
3. Benavent, E., Landete, M., Mota, E., Tirado, G.: The multiple vehicle pickup and delivery problem with LIFO constraints. Eur. J. Oper. Res. **243**(3), 752–762 (2015). http://linkinghub.elsevier.com/retrieve/pii/S0377221714010479
4. Berbeglia, G., Cordeau, J.F., Gribkovskaia, I., Laporte, G.: Static pickup and delivery problems: a classification scheme and survey. Top **15**(1), 1–31 (2007). http://www.springerlink.com/index/10.1007/s11750-007-0009-0
5. Berbeglia, G., Cordeau, J.F., Laporte, G.: Dynamic pickup and delivery problems. Eur. J. Oper. Res. **202**(1), 8–15 (2010)
6. Bräysy, O.: A reactive variable neighborhood search for the vehicle-routing problem with time windows. INFORMS J. Comput. **15**(4), 347–368 (2003)
7. Bräysy, O., Gendreau, M.: Vehicle routing problem with time windows, part I: route construction and local search algorithms. Transp. Sci. **39**(1), 104–118 (2005). http://transci.journal.informs.org/cgi/doi/10.1287/trsc.1030.0056
8. Bräysy, O., Gendreau, M.: Vehicle routing problem with time windows part, II: metaheuristics. Transp. Sci. **39**(1), 119–139 (2005). http://transci.journal.informs.org/cgi/doi/10.1287/trsc.1030.0057
9. Cherkesly, M., Desaulniers, G., Laporte, G.: Branch-price-and-cut algorithms for the pickup and delivery problem with time windows and LIFO loading. Comput. Oper. Res. **62**(1), 23–35 (2015). http://dx.doi.org/10.1016/j.cor.2015.04.002
10. Clarke, G., Wright, J.W.: Scheduling of vehicles from a central depot to a number of delivery points. Oper. Res. **12**(4), 568–581 (1964)

11. Crainic, T.G., Nguyen, P.K., Toulouse, M.: Synchronized multi-tripmulti-traffic pickup & delivery in city logistics. CIRRELT **5**, 1–24 (2015)
12. Demir, E., Bekta, T., Laporte, G.: A review of recent research on green road freight transportation. Eur. J. Oper. Res. **237**(3), 775–793 (2014)
13. Desaulniers, G., Desrosiers, J., Solomon, M.M., Erdmann, A., Soumis, F.: VRP with pickup and delivery. In: Toth, P., Vigo, D. (eds.) The vehicle routing problem, pp. 225–242. SIAM (2002)
14. Dff International Ltd, R.: RHA Cost Tables (2014). http://www.rha.uk.net/, http://www.rha.uk.net/docs/CostTables2014EDITION.pdf
15. Dff International Ltd, R.: RHA National Directory of Hauliers (2015). http://www.rha.uk.net/, https://www.findahaulier.co.uk/
16. Dumas, Y., Desrosiers, J., Soumis, F.: The pickup and delivery problem with time windows. Eur. J. Oper. Res. **54**(1), 7–22 (1991). http://www.sciencedirect.com/science/article/pii/037722179190319Q
17. Gendreau, M., Guertin, F., Potvin, J.Y., Séguin, R.: Neighborhood search heuristics for a dynamic vehicle dispatching problem with pick-ups and deliveries. Transp. Res. part C: Emerg. Technol. **14**(3), 157–174 (2006). http://linkinghub.elsevier.com/retrieve/pii/S0968090X06000349
18. Gendreau, M., Hertz, A., Laporte, G.: New insertion and post optimization procedures for the traveling salesman problem. Oper. Res. **40**(6), 1086–1095 (1992)
19. Gschwind, T., Irnich, S., Mainz, D.: Effective Handling of Dynamic Time Windows and Synchronization with Precedences for Exact Vehicle Routing. Technical report, Johannes Gutenberg University Mainz, Mainz, Germany (2012). http://logistik.bwl.uni-mainz.de/
20. Laporte, G.: Fifty years of vehicle routing. Transp. Sci. **43**(4), 408–416 (2009). http://transci.journal.informs.org/cgi/doi/10.1287/trsc.1090.0301
21. McLeod, F., Cherrett, T., Shingleton, D., Bekta, T., Speed, C., Davies, N., Dickinson, J., Norgate, S.: Sixth Sense Logistics: Challenges in supporting more flexible, human-centric scheduling in the service sector. In: Annual Logistics Research Network (LRN) Conference. Cranfield, UK (2012)
22. Mitrović-Minić, S., Krishnamurti, R., Laporte, G.: Double-horizon based heuristics for the dynamic pickup and delivery problem with time windows. Transp. Res. part B: Methodological **38**(8), 669–685 (2004)
23. Mourdjis, P.J., Cowling, P.I., Robinson, M.: Metaheuristics for the pick-up and delivery problem with contracted orders. In: Blum, C., Ochoa, G. (eds.) Evolutionary Computation in Combinatorial Optimization, pp. 170–181. Springer-Verlag, Heidelberg (2014)
24. Nahum, O.E.: The Real-Time Multi-Objective Vehicle Routing Problem. Ph.d., Bar-Ilan University (2013). http://orennahum.dyndns.org/Files/PhD.pdf
25. Pidd, M.: Computer Simulation in Management Science, 5th edn. Wiley, New York (1998). http://eprints.lancs.ac.uk/47721/
26. Rochat, Y., Taillard, É.D.: Probabilistic diversification and intensification in local search for vehicle routing. J. Heuristics **1**(1), 147–167 (1995)
27. Savelsbergh, M.W.P.: The vehicle routing problem with time windows: minimizing route duration. INFORMS J. Comput. 4(2), 146–154 (1992)
28. Taillard, É.D., Badeau, P., Gendreau, M., Guertin, F., Potvin, J.Y.: A tabu search heuristic for the vehicle routing problem with soft time windows. Transp. Sci. **31**(2), 170–186 (1997). http://transci.journal.informs.org/content/31/2/170.short

29. Toth, P., Vigo, D.: Heuristic algorithms for the handicapped persons transportation problem. Transp. Sci. **31**(1), 60–71 (1997)
30. Xu, H., Chen, Z.L., Rajagopal, S., Arunapuram, S.: Solving a practical pickup and delivery problem. Transp. Sci. **37**(3), 347–364 (2003)

Exploring Techniques to Improve Large-Scale Drainage System Maintenance Scheduling Using a Risk Driven Model

Yujie Chen[1(✉)], Fiona Polack[1], Peter Cowling[1], Philip Mourdjis[1], and Stephen Remde[2]

[1] YCCSA, Computer Science Department, University of York, York, UK
yujiechen369@163.com, {fiona.polack,peter.cowling,pjm515}@york.ac.uk
[2] Gaist Solutions Limited, Lancaster, UK
stephen.remde@gaist.co.uk

Abstract. Gully pots are part of the infrastructure of a storm drain system, designed to drain surface water from streets. Any broken or blocked gully pot represents a potential cause of flooding, for instance during periods of intense or prolonged rainfall. Regular cleaning is necessary for gully pots to function effectively. We model gully pot maintenance as a risk-driven problem, and evaluate the maintenance quality of maintenance strategies by considering the risk impact of gully pot failure and failure behaviour. The results suggest investment directions and management policies that potentially improve the efficiency of maintenance. We find that the current maintenance quality is significantly affected by untimely system status information. We propose that low-cost sensor techniques might be able to improve timeliness of status information, and use simulation results to show the behaviour and advantages that might arise in a range of real-world scenarios.

Keywords: Simulation · Gully-pot system maintenance · Risk management

1 Introduction

Gully pots are designed to prevent solids and sediment from flushing into sewers and causing blockages in the underground system [1]. Regular cleaning is required for gully pots to function effectively [2,3]. Usually, gully pots in a city are cleaned once or twice a year. Partial or complete blockage of the gully pots increases the likelihood of surface water flooding. In extreme situations such as intensive rainfall, a clogged drainage system may cause serious property loss (e.g. [4–7]).

Our gully pot maintenance problem is based on Blackpool, UK. Blackpool's gully pot maintenance system records 28,149 gullies in an area of about 36.1 km^2. On each day, the local authority maintenance team either carries out routine gully pot cleaning, categorised as the preventative maintenance, or responds to

© Springer International Publishing AG 2017
B. Vitoriano and G.H. Parlier (Eds.): ICORES 2016, CCIS 695, pp. 161–179, 2017.
DOI: 10.1007/978-3-319-53982-9_10

emerging events such as gully broken and blockage reports (referred to as corrective maintenance). Depending on the local risk, response to emerging events should be scheduled 5 to 20 days from when they are recorded. For broken gully pots, a specialist vehicle is required. Due to limited human resource, only one vehicle works each day.

Each day there is a schedule of gully pots to visit, starting and ending at the depot. The maintenance vehicle departs the depot at 09:00 and returns no later than 17:00. During servicing, some gully pots are inaccessible due to parked vehicles. Historical maintenance records show that this is a striking issue: about 8.3% of gully pots are not serviced each year because of parked cars.

Apart from the parking issue, we also notice another weakness of current maintenance scheduling strategy – untimely system status information. Currently, all the broken or blocked gully pots are either reported by local residents or found through preventative maintenance. Historically, the records show that reporting of gully pot issues by local residents is highest in autumn, when leaf-fall and higher rain causes many blockages; and lowest in winter, when short daylight and cold weather reduce footfall. This passive situation potentially leads to uncontrolled surface water flooding.

In order to discover techniques or policies that could improve current gully pot maintenance, this paper considers the gully pot maintenance as a risk-driven problem. In our analysis, we take account of each gully pot's failure behaviour and the risk impact of its failure, which varies across the city. The current widely used maintenance strategy, including both preventative and corrective actions, is evaluated by our risk model across various scenarios.

The remainder of this paper is organized as follows. Section 2 reviews maintenance techniques and concept. We then introduce our simulation in Sect. 3. Section 4 shows our results and conclusions. A summary of investment suggestions based on our simulation is provided in Sect. 5.

2   Related Works

Maintenance is generally categorised into corrective and preventative maintenance [8,9]. Corrective maintenance (CM) usually happens after failures occur. It includes actions such as repair and replacement. Preventative maintenance (PM) is an alternative strategy. In industry, preventative maintenance typically takes place at regular time intervals, based on experience.

Operational research on PM introduces decision making, based on data analysis, with techniques such as time-based (TBM) (e.g. [10,11]) and condition-based maintenance (CBM) (e.g. [12,13]). TBM can be applied when the failure rate is predictable, whilst CBM is employed where conditions are continuously monitored by sensors or any appropriate indicators. A similar approach, tracking real-time operation information, is also applied in dynamic scheduling (e.g. [14]). There is a little research combining PM and CM strategies: [15] introduce a PM/CM rate control strategy, obtaining a near-optimal maintenance policy for a manufacturing system.

For TBM, the accurate prediction of the current and future condition of a system is crucial in developing appropriate maintenance schedules. Damage, deterioration and degradation are important notions in asset life cycle management. Literature shows that related research has been done in bridge, pavement and water pipe systems [16–18]. Two techniques are normally applied: first, functional based models, such as exponential [19] or time-powered models [20], have been used to determine the optimal timing of water pipe inspection and replacement; time-dependent Poisson [21] and the accelerated Weibull hazard models [22] are also widely used. Second, Markov chain-based deterioration models have been well studied and applied in a number of real-world applications (e.g. [16–18]). Different from the functional based models, Markov chain-based models focus on the transition probabilities between different grades, which also implies that conditions are evaluated discretely. The advantage of discrete methods is that clear management policies can be addressed based on the corresponding states.

The gully maintenance problem is also related to the periodic vehicle routing problem (PVRP) [23], which is widely used in geographically distributed maintenance and on-site service applications (e.g. [24–27]). The aim is to produce efficient schedule and daily routes that satisfy maintenance frequency and pattern constraints in a given period. However, unlike the above maintenance research, PVRP is based on the assumption that the optimal maintenance frequency and pattern for each object is known.

3   Simulation

3.1   Model for Schedule Strategy in the Real World

Gully pot maintenance is a large-scale problem, and gully conditions change continuously over time. In the real world, a schedule plan is normally provided for the near future (e.g. one week or one month). Therefore, during the planning period, not all gully pots can be serviced.

In order to discover techniques or policies that could improve the current gully pot maintenance, we would like to simulate the actual scheduling strategy that is widely applied across local authorities. We summarise the procedure as follows.

1. Construct efficient preventative maintenance routes. In our model, This subproblem is considered as a vehicle routing problem (VRP). The objective is to build daily cleaning routes that minimize the total travelling distance, with constraints including: (1) all gully pots in the system should be visited at least once; (2) all routes should start and end at the depot; (3) no route travelling time should exceed the working hours constraint. A variable neighbourhood search [28] is applied. A similar solver is described in [29].
2. Collect recent information on emerging broken/blocked gully pots.
3. Generate a maintenance schedule for the near future (e.g. one week or one month). Priority is given to broken and blocked gully reports (corrective maintenance). When all the reported problematic gully pots have been serviced,

the crew comes back to preventative maintenance. To schedule the preventative actions, we give priority to the routes with the highest risk estimates (described in the following section, function 1) that have not been scheduled in the last year.

3.2 Evaluation

In order to evaluate the performance of a maintenance schedule, we propose a risk-driven model. Each day, the risk of surface water flooding due to blocked/broken gully pots is evaluated by function 1:

$$\sum_{i=1}^{N} r_i P_i(d) \tag{1}$$

where N is the total number of gully pots in the drainage system, r_i is the risk impact of gully pot i estimated by its surrounding environment, and $P_i(d)$ is the probability that gully pot i is failed on day d.

The Risk Impact per Gully Pot (r_i). The impact of a hazard such as surface water flooding could be exacerbated by social-related factors, which are usually influenced by economic, demographic and building types [30]. A higher · risk impact here implies that if a particular gully pot is blocked and floods happen, it results in relatively larger economic and social losses. Co-operating with Blackpool local council, we firstly decide a list of social concerns with awareness of their economic and population influence. Then, each gully pot is evaluated by its location and the related social concerns.

Here, social concerns are classified in to three groups: (1) residential property; (2) commercial and industrial areas including local and district centres, business zones, and employment sites; (3) public services including schools, hospitals, doctors and public transport routes. In Table 1, the estimated value of the item in group 1 is the average residential house price in Blackpool [31]. Group 2 takes account of the footfall and critical building prices for each item. The estimated value of items in group 3 is based on average daily operation costs.

Flooding impact analysis involves large uncertainties. We do not expect a precise assessment of impact. Instead, we aim to find values that are able to guide gully pot maintenance actions in decision making. Here, we mainly focus on direct economic losses using a damage function which relates to property type and water level. [32] propose the impact from a range of flood water levels on different building types. After consulting the UK Environment Agency and Blackpool Council, we focus on the impact of flood water levels of less than 21 cm. This gives the value-loss figures shown in Table 1. For public transport we focus on bus routes, estimating the cost of road section closure due to surface water flooding.

By analysing Blackpool's historic flooding frequency [33], the probability of flooding events is used to map the flooding value loss to the daily risk impact

Table 1. Social factor evaluation.

Group	Social concerns	Estimated value	Value loss from flooding	Risk impact
1	Residential	£113,000	3%	£34
2	Local center	£1,130,000	5%	£580
	District center	£1,695,000	5%	£870
	Business area	£565,000	5%	£290
	Employment sites	£226,000	5%	£116
3	School	£5,168	4%	£71
	Large hospital	£917,808	4%	£377
	Doctors	£9,178	4%	£73
	Bus route	£220	100%	£37

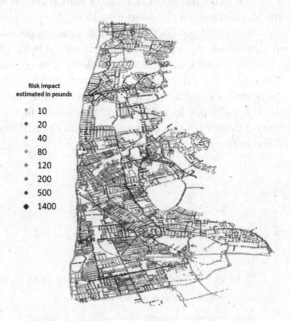

Risk impact estimated in pounds

* 10
* 20
* 40
* 80
* 120
* 200
* 500
◆ 1400

Fig. 1. Gully pot risk impact in Blackpool.

per gully pot according to its location (last column of Table 1). We assume that gullies in the same section of a street evenly share the responsibility for the risk impact evaluated in that area. Figure 1 illustrates the geographic distribution of gully pot risk impact in Blackpool.

Estimating the Process of a Gully Pot Blocking. Ahmad and Kamaruddin [9] suggest that time-based maintenance is the normal strategy in situations where equipment has a fixed lifespan or predictable failure behaviour. After analysis of historic gully pot records, we model the gully pot blocking process using the Weibull distribution model [34,35], from reliability theory. The parameters of this form of Weibull distribution are the scale parameter λ, and the shape parameter k. In our study, all values applied are based on our statistical analysis of the Blackpool data. We first define $k = 6$, which captures a realistically increasing blocking rate over time. The scale parameter λ, capturing lifetime behaviour, is affected by location and seasonal factors, according to a simple linear function:

$$\lambda = \begin{cases} 10 & \text{... if gully pot recorded as broken} \\ E_{calling} & \text{... a calling event} \\ max(90, E - \sum_{f \in F} n_f * s_f) & \text{... normal state} \end{cases}$$

$E_{calling}$ represents the expected number of days from a report on a gully pot to its servicing. E is the expected number of days that it would take a normal gully pot to become blocked since its last services. Here, $E = 10.3$ years. F is a set of factors that may affect gully pot lifetime, such as street type, number of trees nearby, and blown sand effect: n_f represents the effect level from a specific factor $f \in F$ to a gully pot; s_f adjusts the effect from factor f according to seasonal information. For example, if a gully pot is on a street with five deciduous trees nearby, then $n_f = 5$ with $s_f = 93, 1, 389, 433$ in spring, summer, autumn and winter respectively. If a gully pot location is not affected by factor f, we simply assign $n_f = 0$. Figure 2 illustrates one example of a gully pot lifetime estimation taking account of the surrounding environment.

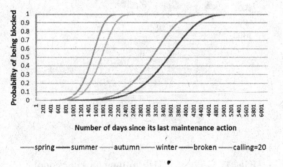

Fig. 2. Probability of being blocked since last maintenance action (Example of a gully pot lifetime with 5 tree nearby at different seasons).

4   Experiment

In this section, we firstly summarise the background of our problem and simulation. All simulations were implemented in C# and executed on a cluster composed of 8 Windows computers, each with 8 cores, Intel Xeon E3-1230 CPU and 16GB RAM.

Simulation Settings

1. Total number of gully pots in the system: 28,149.
2. Broken events: Blackpool council estimates about 1.1% of gully pots are broken every year. This is represented by each gully pot becoming broken randomly with probability $p_b = 0.00003$ per day in our simulation.
3. Blocking probability: a gully pot lifetime is estimated by a Weibull distribution described in Sect. 3.2. Every day, each gully pot has a probability of becoming blocked according to its failure rate function $h_i(d) = \frac{R_i(d-1)-R_i(d)}{R_i(d-1)}$, where $R_i(d) = 1 - F_i(d)$ is the reliability function.
4. Seasonal factors F: the Blackpool data only allows us to include trees and leaf-fall in our simulation. Seasonal factors related to the number of trees nearby highly affect the lifetime of gully pots, and on average, each gully pot is affected by 0.4 trees in Blackpool.
5. Resident calling behaviour: about 1700 calls are received every year by the Blackpool gully maintenance team, and most of the calls concern blocked or damaged gully pots. Over 50 % of all calls occur during the autumn, as shown in Fig. 3. Our statistical analysis determined that, to match the resident calling behaviour in our simulation, on any given day, the probability of receiving a call if a gully pot is already broken or blocked is $p_{calls}(i) = \{0.0033, 0.005, 0.0056, 0.002\}$ for spring through winter, respectively. If a gully pot is not broken, there is still a small chance that a call is received, related to its current condition. The simulation probability is $p_{calls}(i) = P_i(d) * \gamma$, where $\gamma = 10.62$ has been measured experimentally to adjust the calling probability to match the real data.

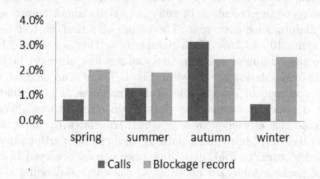

Fig. 3. Seasonal calls and blockages as a percentage of the total number of gully pots in Blackpool.

Simulation Assumption

1. Planning Horizon: In the real world, maintenance schedules are generated at varying levels of granularity, from long term (yearly) to short term (weekly).

Here, we only consider the procedure described in Sect. 3.1, where the maintenance schedule is updated every week according to the most recent system status reports. In addition, the number of working days varies depending on local council requirements: in the following experiments, we assume seven working days per week; holidays are not considered.

2. Parking Issues: inaccessibility during maintenance due to parking predominantly affects preventative maintenance. For corrective actions, including servicing for both resident reports and broken gully pots, our simulation assumes that the team always has access.

3. As well as broken gullies reported by residents, damage is also found during preventative maintenance. In this case, the simulation registers the broken gully and schedules it on a later day.

These parameters and assumption have been discussed with Gaist Solutions Ltd. and agreed to be a realistic representation of gully-pot behaviour in Blackpool.

4.1   The Impact of Parking Issues

According to the maintenance records, the parking issue has been identified as a major problem that decreases the maintenance working efficiency, especially in the old town, where no extra space was designed for parked cars. The number of parked private vehicle also increases significantly. Our simulation helps us to understand the impact of parking on gully-pot maintenance performance. Therefore, potential strategies can be proposed such as banning parking when a maintenance visit for a certain street is scheduled.

In simulation, we can test the effect of inaccessible gully pots using a parameter, x, to represent the percentage of gully pots that cannot be accessed during preventative maintenance each year. The values of x that we test are 0, 5, 8.3 (the actual value), 10 and 15%. Each parameter setting is run over 4 simulated years, with corresponding seasonal factors and residential report behaviours.

The results of simulation are shown in Fig. 4. There is an increase in flooding risk as the percentage of inaccessible drains increases. The simulation result suggests that a policy of suspending parking on streets to be serviced might improve maintenance efficiency by 8%, which translates to about £1,400 risk decrease every day. If a "suspending parking" policy only partially decreases the number of parked cars (to 5%), little difference can be observed in risk. When the impact of parking increases up to 15%, the surface flooding risk increase significantly by 12%.

4.2   What If We Could Do Condition-Based Maintenance (CBM)?

Aside from parking issues, seasonal changes and untimely system status information are identified as other factors that affect the efficiency of drainage system maintenance. Seasonal change is an uncontrollable factor. On the other hand,

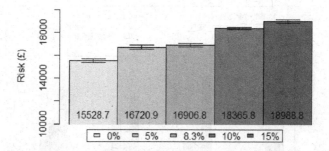

Fig. 4. The average daily risk of applying maintenance schedule described in Sect. 3.1, with different accessibility settings during preventative maintenance. The bar with the setting of 8.3% is the current real-world situation. Error bars show 95% confidence interval.

improving low-cost sensor techniques makes it potentially feasible to continuously monitor gully-pot condition. This would allow our schedule strategies to be combined with CBM, discussed in Sect. 2. Currently, we only find out that a gully pot is blocked or broken either during preventative maintenance or if it is reported; because of this incomplete system information, it is difficult to produce any optimal schedules.

In simulation, we can test the importance of real time failure monitoring by varying the proportion of gully pot failures that are known immediately, as if the gully pot had a real-time sensor. As shown in Table 2, we use two parameters, "since last maintenance action θ" and "percentage of broken gully pots" to control the system's initial state. The stable state assumes that the entire system is well maintained and the number of days since the last maintenance action for each gully is uniformly distributed across 1.1 years. Furthermore, there are about 0.4% broken gullies in the system when it is in the stable situation. The other two scenarios assume that the system is recovering from a natural disaster such that a large number of gullies are broken or blocked initially regardless of prior maintenance. Both a well maintained gully-pot system (see Table 2, recover-1) and a system that has had bad maintenance (see Table 2, recover-2) are tested.

Table 2. "since last maintenance" and "percentage of broken gully pots" set the system's initial state: for all gully pots, the days since their last service are evenly distributed in θ years. We randomly assign a percentage of gully pots to be in the broken state.

	Stable	Recover 1	Recover 2
Since last maintenance θ	1.1	1.1	3
Initial broken gully pots	0.4%	2%	2%

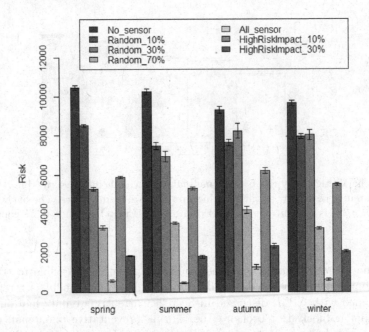

Fig. 5. Performance of maintenance in stable scenario with sensors of different installation capacity. Error bars show 95% confidence intervals. (*Note: This figure corrects
the corresponding results presented in conference proceeding* [36]*, which were generated
using different environment simulation settings (i.e. broken probability* $p_b = 0.00005$,
$\theta = 1.5$ *and initial broken percentage equals 0.7%). However, the analysis and conclusion still holds*).

Figure 5 presents the average daily risk in four seasons over a set of four-year
simulations. In comparison to the simulation of current blockage reporting, the
instant information simulation shows a reduction in risk of about 92%. For the
case where all gully pots have instant (sensor) information, the results clearly
show the impact of seasonal factors: falling leaves and increasing citizen complaints in autumn increase risk by about two times compared to other seasons.
Interestingly, we observe a very different risk distribution across seasons when no
sensors are installed. This is because the different residents' reporting behaviour
in different seasons strongly affects the response time for broken/blocked gullies.
For example, if a gully pots breaks or blocks in winter, this may be only found
out through preventative maintenance in the next year. Such late responses to
problematic gully pots gradually accumulates risk over time. This result uncovers that depending entirely on reporting by local residents effectively hides the
dangers to the system.

To provide further insight into how the availability of information on gully
pots affects flooding risk, we adapt the simulation to provide instant information
from only some locations, simulating the localised installation of sensors. Setting
10% of gullies to have sensors allows us to compare an even distribution of sensors

to the results when sensors are focused on critical areas of the city. We find that focusing on high risk areas reduces the daily risk, on average, by about 28%. When monitoring is increased to cover 30% gullies, the comparable saving is a 75% risk decrease.

Figure 6 illustrates the daily risk change over two years in recovery states. In scenario recovery-1, the system with full sensoring performs the best in terms of recovery speed, followed by the 30% high-risk-impact and the 70% random strategies. The faster recovery also implies lower total surface water flooding risk through the recovery period. In scenario recovery-2, due to the previous poor system maintenance, the recovery period is significantly longer in all cases compared to scenario 1. Also, the peak point uncovers the vulnerability of a

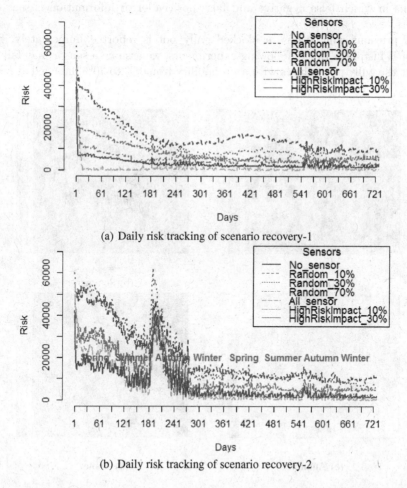

(a) Daily risk tracking of scenario recovery-1

(b) Daily risk tracking of scenario recovery-2

Fig. 6. Performance of maintenance in recovery state with sensors of different install capacity.

badly maintained system during the high-risk season. However, the sensoring still helps the maintenance team to produce a more informed schedule, which results in less total risk during the recovery period.

Reliability. The above simulations show the contribution of timely information to improving the gully-pot system maintenance quality. However, the proposed sensor system also increases the management complexity, where extra cost and manpower are needed to ensure that the system is always working correctly. Furthermore, we assume in our simulation that instant gully-pot condition information can be received with no errors, which is hypothetical. In practice, current sensor techniques can achieve up to 85% reliability [37]. To justify the benefit from potential sensor technique in more realistic scenarios, we test various situations in which false negative and false positive error information is sent by sensors.

In previous simulations, a blocked gully pot is reported immediately if a sensor is installed. In the following experiments, we assume a sensor may fail to report the gully pot failures with a probability from 0% to 30%, labelled as false

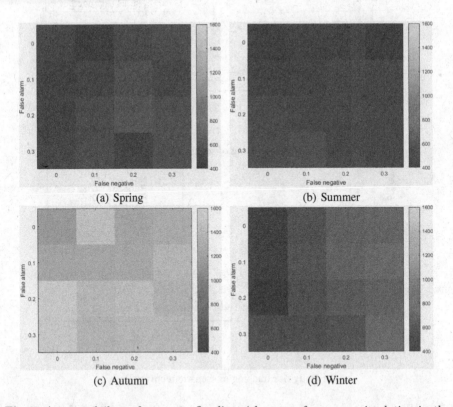

Fig. 7. Average daily surface water flooding risk over a four years simulation in the stable scenario. A full sensoring system with false positive and false negative errors is considered.

negative error in Fig. 7. In this situation, the gully pot failure information relies on traditional reporting from local residents. Meanwhile, for any gully pot in its normal state, an installed sensor may send a false alarm with probability from 0% to 30%. We run the maintenance simulation with a full sensoring system over four years in the stable scenario (see Fig. 5).

The results of average daily surface water flooding risk are illustrated in Fig. 7. From Figs. 5 and 7, we can see that the overall maintenance quality in a full sensoring system, in terms of surface water flooding risk management, is still well controlled across all seasons even with the probability of both false positive and false negative errors up to 30%. In the relatively stable seasons (spring and summer), the false reports show no strong affects on maintenance quality. In the autumn period, many reactive actions emerge due to the increasing failure rate of each gully pot and the number of local residents' complaints. A large number of false alarms significantly disrupts necessary maintenance in this period, resulting in a risk increase of about 33%. In the winter period, when the failure rate of each gully pot is high but residents' reports are rare, there is a clear trend: higher false negative error leads to worse risk. This result reveals the importance of a reliable sensoring system, particularly if it is the dominate information resource.

Further Discussion. Whilst the use of sensors might be of benefit in maintenance scheduling and risk reduction, the realism of this approach needs further consideration. Accurate sensor information depends not just on the sensor detecting problems, but also on communication performance, which decreases in weather condition such as rain or snow [37]. The gully-pot system maintenance should combine a risk estimation approach (i.e. Sect. 3.2) with sensors to deliver optimized scheduling.

Our simulation shows large advantages when sensors are installed in high-risk areas. However, since sensors must be close enough to communicate wirelessly with each other, optimisation of the sensor network topology must be considered [37, 38].

4.3   Can We Reduce Maintenance Frequency When Providing CBM?

Historic gully pot maintenance records from several local councils of the UK, show that the working frequency and pattern vary according to local policies. However, our simulation experiments so far have assumed so far that the maintenance schedule is updated every week and the crew works 7 days a week (see Sect. 4). In order to further explore whether the installation of sensors is worthwhile, we compare the impact of reducing maintenance frequency on a no-sensoring system and full-sensoring system. To set up the simulations, the same scheduling policy is used (see Sect. 3.1), except that the maintenance crew only works for the first x days every week.

Figure 8 shows the average daily risk over a four-year simulation with different working frequency settings. Firstly, we can see the that the risk increases

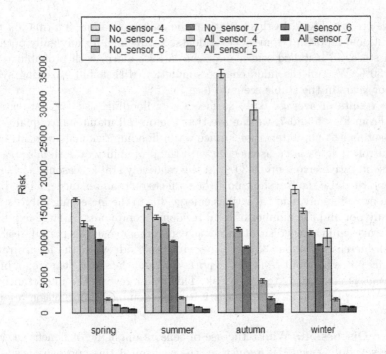

Fig. 8. Performance of maintenance from a stable state with different working frequency: from four days per week to seven days per week. Error bars show 95% confidence intervals.

while the working frequency decreases in both no-sensoring and sensoring system. In the relatively low risk seasons of spring and summer, the advantage from using a sensoring system is still apparent when we reduce the working frequency down to four days per week. However, the lack of maintenance leads to a series of problems in high risk seasons (autumn and winter). Our results suggest that local authorities might make savings by using different working frequencies according to seasonal information. Comparing the result of five working days with sensoring to seven working days without sensoring, we can potentially improve the maintenance quality by about 78%, whilst reducing working time by 30%.

Figure 9 illustrates detailed daily risk changes over five-year simulations under various working frequencies. All simulations start from the same initial state as the stable scenario shown in Table 2, which assumes previous working frequency is seven days per week. Firstly, we can see that, when applying sensoring, the gully-pot system can be restored within about a month. In the first year, the risk fluctuates at about 600 and 10,000, for the system with and without sensoring respectively. A dramatic risk increase can be observed in the second autumn period when we decrease the working frequency down to four days per week in both no-sensoring and sensoring system, and this risk fluctuation pattern repeats in subsequent autumns. Once again, the result shows that insufficient maintenance will lead the system to a vulnerable situation, especially

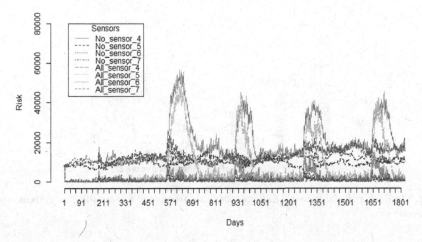

Fig. 9. Performance of maintenance from a stable state with different working frequency over five years simulations.

during high risk seasons. Furthermore, this effect persists, even when the perfect system status information is given by sensors.

The results above illustrate the minimum number of days (i.e. five days a week) required to maintain the advantage due to timely system status information. To further explore the damage of insufficient maintenance frequency, we decrease the number of working days down to three days per week. The results, Fig. 10, show that there is no significant effect from this reduction of working frequency in the first year. However, the reduced maintenance has resulted in gradually accumulated hazards, and these are suddenly exposed later, in what appears to be a critical threshold effect.

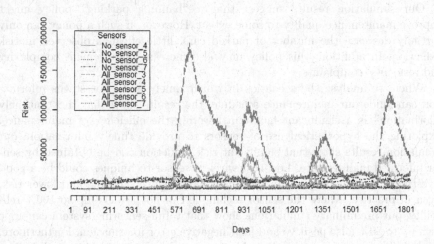

Fig. 10. Performance of maintenance from a stable state with different working frequency over five years simulations.

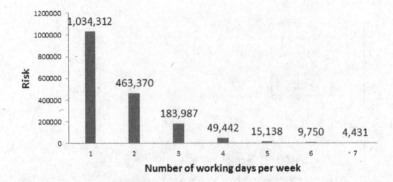

Fig. 11. Peak risk value using different working frequency for a gully-pot system with sensors.

Focusing on the peak values, Fig. 11 plots the worst risk value found in the simulation runs for the different working frequencies. The results show a clear, exponential risk increase in the worst situation when we decrease the working frequency every week. This result again sets up an alert of potential consequences when a system lacks sufficient maintenance.

5   Conclusion

This paper considers a real-world drainage system maintenance problem. A risk-driven analysis approach is proposed to evaluate the performance of maintenance actions. We have used simulation to explore the effect on maintenance scheduling and risk of "parking issues" and "untimely system status information", both of which are known weaknesses of the current maintenance approach (see Sect. 3.1).

Our simulation results suggest that a "banning parking" policy might improve maintenance quality to some extent. However, if such a policy can only partially decrease the number of parked cars, little effect is observed in risk reduction. In addition, this policy may also increase management complexity and residents' complaints.

When we analyse the scenarios in which timely gully pot status information can guide our maintenance schedule, the results show that the "untimely information" is a significant factor in lowering the efficiency of maintenance. Exploring the hypothetical use of sensors to provide timely information, our simulation results show that significant risk reduction can be obtained by sensor informed maintenance. Low-cost wireless sensor techniques could be a good investment to help produce an informed maintenance schedule and lower risk. Even in practice, where the sensor technique can not achieve up to 100% reliability, our preliminary simulations show that a full-sensing system can cope with up to 30% false positive and false negative error information. Furthermore, the benefit of sensing can still hold when we reduce the working week by two days.

Further work is needed to form a cost/benefit analysis to discover the optimal quantity of sensors to deploy, their locations and network topology. New scheduling approaches may be required to make best use of the potentially large amount of data generated by the sensors.

In practice, due to the immaturity of sensor technology, we suggest that the combination of time-based preventative maintenance (with risk estimation) and condition-based corrective maintenance (with sensors) is an optimal approach.

Acknowledgements. The authors would like to thank Gaist Solutions Ltd. for providing data and domain knowledge. This research is part of the LSCITS project funded by the Engineering and physical sciences research council (EPSRC).

References

1. Butler, D., Xiao, Y., Karunaratne, S.: The gully pot as a physical, chemical and biological reactor. Water Sci. Technol. **31**, 219–228 (1995)
2. Karlsson, K., Viklander, M.: Polycyclic aromatic hydrocarbons (PAH) in water and sediment from gully pots. Water Air Soil Pollut. **188**, 271–282 (2008)
3. Scott, K.: Investigating Sustainable Solutions for Roadside Gully Pot Management. Ph.D. thesis (2012)
4. BBC: Flooding affects Pembrokeshire residents and businesses (2011). http://www.bbc.co.uk/news/uk-wales-15441912. Accessed 04 Sept 2015
5. BBC: Floods: North Wales Police travel warning after rain (2012). http://www.bbc.co.uk/news/uk-wales-19702806. Accessed 05 Sept 2015
6. Shieldsgazette: Flooding hell caused by blocked gully (2012). http://www.shieldsgazette.com/news/local-news/flooding-hell-caused-by-blocked-gully-1-4761698. Accessed 24 Aug 2015
7. Leylandguardian: Blocked drain causes flooding danger on Lancashire road (2015) http://www.leylandguardian.co.uk/news/blocked-drain-causes-flooding-danger-on-lancashire-road-1-7425326. Accessed 24 Aug 2015
8. Duffuaa, S., Ben-Daya, M., Al-Sultan, K., Andijani, A.A.: A generic conceptual simulation model for maintenance systems. J. Q. Maintenance Eng. **7**, 207–219 (2001)
9. Ahmad, R., Kamaruddin, S.: An overview of time-based and condition-based maintenance in industrial application. Comput. Ind. Eng. **63**, 135–149 (2012)
10. Scarf, P.A., Cavalcante, C.A.V.: Hybrid block replacement and inspection policies for a multi-component system with heterogeneous component lives. Eur. J. Oper. Res. **206**, 384–394 (2010)
11. Wu, J., Adam Ng, T.S., Xie, M., Huang, H.Z.: Analysis of maintenance policies for finite life-cycle multi-state systems. Comput. Ind. Eng. **59**, 638–646 (2010)
12. Carnero Moya, M.C.: The control of the setting up of a predictive maintenance programme using a system of indicators. Omega **32**, 57–75 (2004)
13. Campos, J.: Development in the application of ICT in condition monitoring and maintenance. Comput. Ind. **60**, 1–20 (2009)
14. Cowling, P., Johansson, M.: Using real time information for effective dynamic scheduling. Eur. J. Oper. Res. **139**, 230–244 (2002)
15. Kenne, J.P., Nkeungoue, L.J.: Simultaneous control of production, preventive and corrective maintenance rates of a failure-prone manufacturing system. Appl. Numer. Math. **58**, 180–194 (2008)

16. Madanat, S., Ibrahim, W.H.W.: Poisson regression models of infrastructure transition probabilities. J. Transp. Eng. **121**, 267–272 (1995)
17. Morcous, G., Rivard, H., Hanna, A.M.: Modeling bridge deterioration using case-based reasoning. J. Infrastruct. Syst. **8**(3), 86–95 (2002)
18. Baik, H.S., Jeong, H.S.D., Abraham, D.M.: Estimating transition probabilities in Markov chain-based deterioration models for management of wastewater systems. J. Water Resour. Planning Manage. **132**, 15–24 (2006)
19. Shamir, U., Howard, C.D.: An analytic approach to scheduling pipe replacement. J. Am. Water Works Assoc. **71**, 248–258 (1978)
20. Kleiner, Y., Rajani, B.: Comprehensive review of structural deterioration of water mains: statistical models. Urban Water **3**, 131–150 (2001)
21. Constantine, G., Darroch, J., Miller, R.: Predicting underground pipeline failure. WATER -MELBOURNE THEN ARTARMON **23**, 9–10 (1996)
22. Le Gat, Y., Eisenbeis, P.: Using maintenance records to forecast failures in water networks. Urban Water **2**, 173–181 (2000)
23. Christofides, N., Beasley, J.E.: The period routing problem. Networks **14**, 237–256 (1984)
24. Shih, L.H., Chang, H.C.: A routing and scheduling system for infectious waste collection. Environ. Model. Assess. **6**, 261–269 (2001)
25. Gaur, V., Fisher, M.L.: A periodic inventory routing problem at a supermarket chain. Oper. Res. **52**, 813–822 (2004)
26. Claassen, G.D.H., Hendriks, T.H.B.: An application of Special Ordered Sets to a periodic milk collection problem. Eur. J. Oper. Res. **180**, 754–769 (2007)
27. An, Y.J., Kim, Y.D., Jeong, B.J., Kim, S.D.: Scheduling healthcare services in a home healthcare system. J. Oper. Res. Soc. **63**, 1589–1599 (2012)
28. Hansen, P., Mladenović, N., Moreno Pérez, J.A.: Variable neighbourhood search: methods and applications. Ann. Oper. Res. **175**, 367–407 (2010)
29. Chen, Y., Cowling, P., Remde, S.: Dynamic period routing for a complex real-world system: a case study in storm drain maintenance. In: Blum, C., Ochoa, G. (eds.) EvoCOP 2014. LNCS, vol. 8600, pp. 109–120. Springer, Heidelberg (2014). doi:10. 1007/978-3-662-44320-0_10
30. Cutter, S.L., Carolina, S., Boruff, B.J.: Social vulnerability to environmental hazards. Soc. Sci. Q. **84**, 242–261 (2003)
31. UK GOV: Price Paid Data 2015 (2015). https://www.gov.uk/government/statistical-data-sets/price-paid-data-downloads. Accessed 09 Sept 2015
32. Thieken, A., Ackermann, V., Elmer, F., Kreibich, H., Kuhlman, B., Kunert, U., Maiwald, H., Merz, B., Muller, M., Piroth, K., Schwarz, J., Schwarze, R., Seifert, I., Seifert, J.: Methods for the evaluation of direct and indirect flood losses (2008)
33. Blackpool: Blackpool strategic flood risk assessment. Technical report, June 2009. https://www.blackpool.gov.uk/Residents/Planning-environment-and-community/Documents/Blackpool-Strategic-Flood-Risk-Assessment.pdf. Accessed 05 July 2015
34. Weibull, W.: A statistical distribution function of wide applicability. J. Appl. Mech. **18**, 293–297 (1951)
35. Ebeling, C.E.: An introduction to reliability and maintainability engineering. Tata McGraw-Hill Education (2004)
36. Chen, Y., Polack, F., Cowling, P., Mourdjis, P., Remde, S.: Risk driven analysis of maintenance for a large-scale drainage system. In: 5th International Conference on Operations Research and Enterprise Systems (ICORES) (2016)

37. See, C.H., Horoshenkov, K.V., Abd-alhameed, R.A., Hu, Y.F., Tait, S.J.: A low power wireless sensor network for gully pot monitoring in urban catchments. IEEE Sens. J. **12**, 1545–1553 (2012)
38. Yick, J., Mukherjee, B., Ghosal, D.: Wireless sensor network survey. Comput. Netw. **52**, 2292–2330 (2008)

A Comparison of One-Pass and Bi-directional Approaches Applied to Large-Scale Road Inspection

Yujie Chen[1(✉)], Fiona Polack[1], Peter Cowling[1], and Stephen Remde[2]

[1] YCCSA, Computer Science Department, University of York, York, UK
yujiechen369@163.com, {fiona.polack,peter.cowling}@york.ac.uk
[2] Gaist Solutions Limited, Lancaster, UK
stephen.remde@gaist.co.uk

Abstract. Gaist Solutions Ltd. carries out national-scale road inspection surveys in the UK. Visual inspection is used to identify the need for road maintenance. An inspection vehicle that monitors one side of the road needs two traversals to monitor a typical road, whereas a vehicle with cameras that record both sides of the road only requires a one-pass approach. To determine whether the one-pass approach affords any real cost advantage, we analyse road networks of six typical UK cities and the county of Norfolk using a range of exact and heuristic methods, and extrapolate from our results to estimate the cost-effectiveness of these two approaches for the road network of the UK. Our analysis approach is based on the Chinese Postman Problem (CPP), using graph reduction to allow effective computation over very large data sets.

Keywords: Routing · Revenue management · Chinese Postman Problem

1 Introduction

In the UK, local government agencies have a duty to maintain roads for public utility and safety. Road inspection is cyclically scheduled and done at a defined frequency, to support decision making on road repair scheduling. Modern inspection uses vehicles equipped for high quality video recording. The quality of recording is affected by obstructions such as parked cars, and complex post-processing is required to extract suitable data on road condition. With 240,000 km of road in the UK, efficiency of the inspection process has high priority. Two branches of research have shown promising results: firstly, using the asset deterioration process in optimising inspection frequency and policies (e.g. [18, 23, 24, 27]); secondly, optimising daily inspection routes with time constraints (e.g. [16, 17]).

We introduce a third option for improving operational efficiency via the road inspection strategies. We investigate two strategies: a one-pass inspection, in which an inspection vehicle can monitor both sides of a road in one traversal, and the more traditional bi-directional inspection, in which the vehicle monitors only its near-side carriageway. We use UK data for the road networks of Blackpool,

© Springer International Publishing AG 2017
B. Vitoriano and G.H. Parlier (Eds.): ICORES 2016, CCIS 695, pp. 180–200, 2017.
DOI: 10.1007/978-3-319-53982-9_11

Southend, Manchester, Stockport, Halton, Warrington and the rural county of Norfolk, with total road lengths of 515, 508, 1315, 945, 619, 879 and 26243 km, respectively.

1.1   Road Networks and Representations

In the UK, local authorities are responsible for local road networks (excluding major trunk roads and motorways). The local authority road networks are mostly designated as 1-lane, 2-lane, 3-lane and 4-lane single carriageways or 2-lane dual carriageways (with a central reservation).

We represent the road network as an undirected graph $G(V, E)$. The vertices V represent junctions, dead ends, bends and any data collection point identified in the original data. The edges E represent roads that link the vertices. As Figs. 1 and 2 show, 1-lane, 2-lane and 3-lane single carriageways may be represented by a single undirected edge. In the 4-lane single carriageway and dual carriageway situations, we transform the road into two parallel undirected edges. A crescent road is transformed into a loop (see Fig. 3) and a cul-de-sac is represented as an one-degree vertex (see Fig. 4).

Our data comprises basic road information from seven UK local councils that are Gaist's clients. Because the data was not collected for network analysis, there are some accidental omissions and other issues to be addressed. To clean the data, all intersections have to be explicitly labelled as vertices.

The pre-processing required for any inspection route analysis is as follows.

– The first change made to the data is to remove 2-degree vertices, since these represent a bend in a road, rather than an intersection, and thus have no

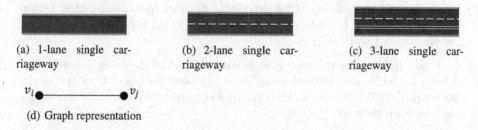

(a) 1-lane single car-
riageway

(b) 2-lane single car-
riageway

(c) 3-lane single car-
riageway

(d) Graph representation

Fig. 1. Common road types in urban areas and the corresponding graph representation.

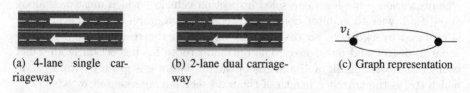

(a) 4-lane single car-
riageway

(b) 2-lane dual carriage-
way

(c) Graph representation

Fig. 2. 4-lane single carriageway, 2-lane dual carriageway and the corresponding graph representation.

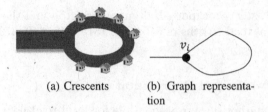

(a) Crescents   (b) Graph representa-
                       tion

Fig. 3. Crescents and the corresponding graph representation.

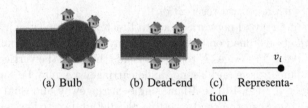

(a) Bulb   (b) Dead-end   (c)   Representa-
                                      tion

Fig. 4. Cul-de-sac and the corresponding graph representation.

impact on the construction or distance of inspection tours. For all two-degree vertices, v_k, $e(v_i, v_k)$ and $e(v_k, v_j)$ are replaced by a single edge, $e(v_i, v_j)$, with length $l((v_i, v_j)) = l((v_i, v_k)) + l((v_k, v_j))$.

- We assume that an intersection has been omitted if the data indicates that two roads terminate close together. Thus, where the distance between two one-degree vertices is smaller than some ε, the vertices are merged. From consideration of the original data, we choose to define $\varepsilon = 3\,\mathrm{m}$.

- Similarly, we assume that a road that terminates very close to another road is an omitted intersection. Thus, where the distance from a one-degree vertex v_k to an edge $e(v_i, v_j)$ is less than ε, the edge $e(v_i, v_j)$ is replaced by two new edges $e(v_i, v_k)$ and $e(v_k, v_j)$.

Table 1 summarises the vertex degree-distribution of the seven datasets that we have available, after data cleaning. As we can see, the majority of vertices in road maps have degree two, so their removal significantly reduces the size of the graph for each network.

1.2   Analysis of Bi-directional and One-Pass Approaches

The inspection route for a one-sided inspection vehicle – which must pass along every road twice to monitor both sides – comprises a graph in which every edge is replaced by two arcs, so that all the vertices in the resulting digraph have equal in-degree and out-degree. The optimum route for inspection is an Euler tour of the resulting digraph and the optimum distance is given by Eq. 1 [8], in which $l(e)$ is the travelled length of the road segment represented by edge e.

$$l(bi\text{-}directional) = 2 \sum_{e \in E} l(e) \tag{1}$$

Table 1. Distribution of vertex degrees, and total number of vertices before and after removal of 2-degree vertices for seven local authority road networks in the UK. Road network from left to right– B: Blackpool; SO: Southend; M: Manchester; ST:Stockport; H:Halton; W:Warrington; N: Norfolk

	B	SO	M	ST	H	W	N
Vertex total	26302	22864	45408	44470	29610	24518	549345
Degree							
1	3.40%	5.60%	8.32%	6.12%	6.08%	10.68%	2.62%
2	80.60%	80.00%	70.63%	80.51%	83.80%	69.53%	91.82%
3	13.10%	12.97%	19.75%	12.31%	9.72%	19.61%	5.33%
4	2.80%	1.42%	1.26%	1.03%	0.36%	0.18%	0.22%
5	0.03%	0.02%	0.04%	0.03%	0.017%	–	–
6	0.01%	–	–	–	–	–	–
Less degree-2s	5103	4571	13337	8665	4796	7471	44912

In the one-pass road inspection case, each edge in the graph has to be visited at least once. This problem can be modelled as the Chinese postman problem (CPP) and the total travel distance is the length of Chinese Postman Tour (CPT) [13, 29].

$$l(one\text{-}pass) = l_{CPT} \tag{2}$$

Since we generate an Eulerian graph from a given road network, Lemma 1 means that we can always find a CPT that passes complementary directions of parallel edges, and which is thus valid for 4-lane single and dual carriageways (e.g. Fig. 2).

Lemma 1. *If a graph $G(V, E)$ is Eulerian and edge $\{v, w\}$ appears twice in E, then there is an Euler tour of G where $\{v, w\}$ is travelled in both directions $\{v, w\}$ and $\{w, v\}$.*

Proof. Suppose that there is an Euler tour $v - w - x_1 - x_2 - ... - x_n - v - w - y_1 - y_2 - ... - y_n - v$, where edge (v, w) is travelled in the same direction both times. Then $v - w - v - y_n - y_{n-1} - ... - y_1 - w - x_1 - x_2 - ...x_n - v$ is another Euler tour where edge (v, w) is travelled in both directions.

One-way streets make up a very small proportion of the total road distance in our networks, so we make the assumption that inspection vehicles can traverse roads in either direction, and that the effect of data-cleaning or data errors is similar in both the one-pass and bi-directional approaches, and thus has minimal effect on our analysis.

1.3   Solution of the Chinese Postman Problem and the Challenge of Large-Scale Problems

The optimal solution to the one-pass inspection problem is a CPT, in which a vehicle must visit every edge at least once whilst travelling the least overall distance [13,29].

For a general undirected graph, a CPT is derivable by adding the smallest possible number of edges to construct an Eulerian graph and finding an Eulerian tour based on it.

Edmonds and Johnson [7] provide a widely-used CPP solution that is polynomial on the number of vertices and edges, as follows:

1. From an undirected graph $G(V, E)$, find the shortest path between all pairs of odd-degree vertices.
2. Find the minimum-cost perfect matching, M, of odd-degree vertices using the blossom algorithm [6,7].
3. Add extra edges that connect all the matched pairs of vertices through the shortest path in G.
4. Find an Eulerian tour in the resulting Eulerian graph.

Then, the length, l_{CPT}, of the identified CPT is:

$$l_{CPT} = \sum_{e \in E} l(e) + l(M) \tag{3}$$

The approach of [7] does not scale well to large graphs, such as our road network representations. The first step of the approach requires calculation of the shortest path in G between every pair of odd-degree vertices. Floyd [10] and others developed an algorithm of complexity $O(n^3)$, where n is the number of vertices, now known as the Floyd-Warshall algorithm (FW), whereas the most efficient implementations of Dijkstra's algorithm [5], a single-source shortest-path algorithm, can achieve $O(m + n \log n)$, where m is the number of edges [11]. There is ongoing debate over the most efficient way to find the all-pairs shortest path in large-scale sparse graphs [28]; tests on our graphs show that FW systematically outperforms the other algorithms.

The second step of the approach by [7], minimum-cost perfect matching, is also computationally expensive. The best-known implementation of the blossom algorithm achieves $O(n(m+n \log n))$ by Gabow [12]. More recently, Kolmogorov [19] published an executable implementation which achieves time complexity of $O(n^2m)$ – again, the complexity of the algorithm is dependent on the number of edges and vertices in the graph. This matching strategy is also used by Christofides [3] for the travelling salesman problem giving a worst-case ratio of $3/2$ of the optimum tour length.

There are several efficient approaches for identification of an Euler tour, required for both one-pass and two-pass inspection routing. Fleury [9] proposes the best-known algorithm, of order $O(m^2)$. However, we use another algorithm of order $O(m)$, proposed by Hierholzer [14].

Apart from Edmonds' CPP solution, Laporte [21] introduces methods of transforming an arc routing problem into an equivalent TSP. This idea is also shown by Irnich [15] to solve a large-scale real-world postman problem with complex constraints. Heuristics for the TSP can then be used to solve the transformed CPP problems. As a result, no optimal result is guaranteed to be found.

To analyse our large-scale real-world road inspection networks, we firstly propose a novel graph reduction process before finding the CPT. Next we compare one-pass and bi-directional inspection strategies for the seven local authority road networks. We also apply our approach to three groups of simulated scenarios. Finally, we present the estimation of using these two inspection strategies on the entire UK road networks.

2   Finding Optimal Inspection Routes

In this section, we describe how we apply the 4-step approach outlined above to our road network graphs. However, our first step is to reduce the graph, to make it more amenable to computation.

2.1   Graph Reduction

Our graph reduction applies graph contraction techniques as used in graph minor theory [2,22]. Edge contraction is a fundamental operation in graph minors which deletes an edge from a graph G and merges the two end points. Here, we propose a novel graph reduction method to decrease the calculations time whilst maintaining the necessary characteristics of the original graph, to allow us to reconstruct a CPT. After the data preparation described in the previous section, each road network is represented as a finite undirected graph which contains parallel edges and self-loops. All degree-2 vertices have been removed.

Let V_{even} and V_{odd} describe the even-degree vertex set and odd-degree vertex set, respectively, of our graph $G(V, E)$. $l(v_i, v_j)$ represents the length of the shortest edge between vertices v_i and v_j. If there is no direct connection between v_i and v_j, $l(v_i, v_j) = \infty$. For all vertices v_i, $l(v_i, v_i) = 0$. The shortest path between v_i and v_j in the graph G is represented as $p(v_i, v_j)$.

Our approach deletes vertices systematically, but records the length and location of removed edges in a structure, E^*. Figure 5 shows how this works on a stylised representation of a road network graph.

We make the following observations.

1. Deleting a self-loop from a graph does not change the parity of a vertex's degree.
2. The shortest path $p(v_i, v_j)$ between vertex v_i and vertex v_j does not include any self-loops.
3. The paths P of the minimum cost matching $M(V_{odd})$ include all the edges connected to degree one vertices. In other words, if you reach a dead end, then you have to get out the same way.

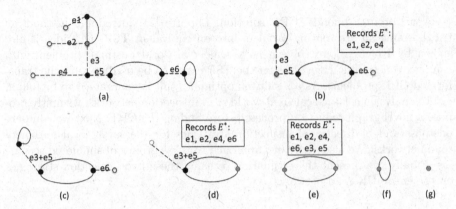

Fig. 5. Systematic reduction of an undirected graph. (a) the graph after removal of degree-2 vertices, with the matching M_{odd} shown in dashed lines. (b) the graph after removal of degree-1 vertices, with their originally connected edges recorded in E^*. (c)-(g) the results of repeating these steps – here, the result is a null graph. White nodes have degree 1; striped nodes have even degree and black nodes have odd degree.

4. If a shortest path between vertex v_i and v_j is $p(v_i, v_j) = (v_i, v_{i+1}, v_{i+2}, ..., v_j)$, then the shortest path between v_{i+1} and v_j is $p(v_{i+1}, v_j) = (v_{i+1}, v_{i+2}, ..., v_j)$.
5. Deleting a degree-1 vertex and its adjacent edge, the total number of odd degree vertices is either unchanged or reduced by 2.

From these observations, the following two deductions can be made.

1: Deleting a self-loop (v_i, v_i) from a graph G will not change the paths in the minimum perfect matching $M(V_{odd})$ of an undirected graph.
2: There is a path set P of the matching $M(V_{odd})$ of the original graph G (as shown by dashed lines in panel (a) of Fig. 5) that equals the deleted edges E^* connected to one-degree vertices (as shown by the E^* in Fig. 5(b)), plus a path set P' in the new matching $M'(V_{odd})$ of the simplified graph G'(shown in Fig. 5(b) as dashed lines), such that, $P = E^* \cup P'$.

Having reduced our road network graphs, we then apply the 4-step process, as follows.

2.2   Step 1: Finding the Shortest Distance Between All Odd-Degree Vertices

Our graph reduction process results in a simplified graph G', and a record of all the deleted edges which were connected to one-degree vertices in the reduction process, E^*.

To find the shortest distance between all pairs of odd-degree vertices of graph G', we use Floyd-Warshall (FW) algorithm [10]. The FW algorithm's complexity is worst-case $O(|V|^3)$, so reducing the number of vertices in the graph is advantageous. We find that, after applying the graph reduction process above, there are still be many even-degree vertices in G', and the matching process does not need these vertices. Therefore, before calculating the shortest path, we can consider deleting these even-degree vertices.

There are many approaches to even-degree vertex deletion. Our preferred approach is Algorithm 1, which detects and deletes all even-degree vertices in G' without affecting the shortest connection and distance between other vertices. Deleting vertices with degree bigger than three may increase calculation complexity and the total number of edges in the graph. However, even where the number of edges increases, the total number of vertices is reduced. The time complexity of deleting each even-degree vertex $v_i \in V_{even}$ is $((m(v_i)(m(v_i) - 1))\,/\,2)$, where $m(v_i)$ is the number of edges connected to vertex v_i.

Having performed the additional reductions using Algorithm 1, we use the FW to calculate the shortest path between all vertices remaining in the graph.

Algorithm 1. Even-degree vertex selection and deletion. V and E are the sets of all vertices and all edges in a given graph G.

---

for each vertex $v_i \in V_{even}$ **do**
   for each pair of edges $e_p(v_k, v_i) \in E$ and $e_q(v_m, v_i) \in E$ **do**
      if $l(e_p) + l(e_q) < l(v_k, v_m)$ **then**
         generate a new edge $e'(v_k, v_m)$ with cost $l(e') = l(e_p) + l(e_q)$
         if $e(v_k, v_m) \in E$ **then**
            replace the edge between v_k and v_m with the new edge $e'(v_k, v_m)$
         else
            add the edge $e'(v_k, v_m)$ between v_k and v_m
         end if
      end if
      delete edges $e_p(v_k, v_i)$ and $e_q(v_m, v_i)$
   end for
   delete vertex v_i
end for

2.3 Step 2: Minimum-Cost Perfect Matching

The standard blossom algorithm [7] finds the minimum-cost perfect matching, $M(V'_{odd})$ of graph G'.

The length of the minimum perfect matchings is represented as $l_{M(V'_{odd})}$. According to Deduction 2 and Eq. 3, the length l_{CPT} of the CPT of the original graph G is:

$$l_{CPT} = \sum_{e \in E} l(e) + \sum_{e \in E^*} l(e) + l_{M(V'_{odd})} \tag{4}$$

2.4 Step 3: Construct the Eulerian Graph

Using Deduction 2 to construct an Eulerian graph from the original graph G, we only need to add edges recorded in E^* and $M(V'_{odd})$ to the original graph G.

2.5   Step 4: Finding the CPT in Original Graph G

From the graph produced in step 3, the Euler tour can be found by applying the algorithm proposed by Hierholzer [14].

To generate a CPT in a real world situation, when Hierholzer's algorithm meets a vertex connected to 4-lane single carriageway or dual carriageway edges, priority is given to the edge whose underlying direction is away from this vertex.

3   Experimental Set-Up and Results

In order to justify the impact of our graph reduction process, we firstly introduce three heuristic methods to solve the CPP. We then run all three approaches plus the 4-step approach of [7] on the road network graphs, firstly without our graph reduction, then on the graphs after our graph reduction method, and finally on the further reduced graph with all even-degree vertices removed before the shortest distance calculation.

Our heuristic approaches focus on the matching process (*Step2*) and retain all the other CPP solving steps of Edmonds' method. Blossom and the first two heuristics labelled *greedy* and GLS (*greedy method with local search*) require the FW calculation of shortest paths between odd-degree vertex pairs. The final heuristic is a breadth-first-search (labelled BFS) for matching, and does not need FW to run first. The combinations of CPP solvers and graphs are labelled $m1 \ldots m12$, as shown in Table 2.

Table 2. Summary of the three graphs and four methods applied. Cells (m1 - m12) provide the key to labelling of the later results and graphs. FW = Floyd-Warshall algorithm to find shortest distance between vertex pairs, step 1. Blossom, greedy, GLS and BFS are tested in the matching process, step 2. GR = Graph reconstruction, step 3. HA = Hierholzer Algorithm to find the CPT, step 4.

Graph	Steps 2–5			
	FW, *blossom*, GR, HA	FW, *Greedy*, GR, HA	FW, *GLS*, GR, HA	*BFS*, GR, HA
Road Network with 2-degree vertices removed	m1	m2	m3	m4
.. and reduction applied (Sect. 2.1)	m5	m6	m7	m8
.. and all even degree vertices removed	m9	m10	m11	m12

The base-case minimum-cost matching is calculated using the implementation by Kolmogorov [19] of the blossom algorithm. We now introduce the three heuristics that are proposed in place of the blossom matching algorithm.

Greedy Method (Greedy): The greedy method is a heuristic that systematically constructs a matching where shortest distances between pairs of vertices are known, as described in Algorithm 2. The algorithm is based on those by Kurtzberg [20] and Reingold et al. [26].

Algorithm 2. Greedy construction for matching.

VL is a list contains all the vertices, v_i in a given graph
while VL contains at least two vertices **do**
    Choose the pair of vertices with shortest distance $(v_i, v_j) \in VL$
    Add (v_i, v_j) to the matching M
    delete v_i, v_j from VL
end while
Return M

Greedy Method + Local Search (GLS): GLS attempts to improve the result of the greedy method by following Algorithm 2 with a greedy first improvement heuristic, shown in Algorithm 3.

Breadth First Search (BFS): BFS is a basic search: each unmatched odd-degree vertex v_i in graph G is the root point of a BFS to find the next unmatched odd-degree vertex v_j. Then two vertices v_j and v_i are a matching pair and the path from v_i to v_j in the BFS tree is the matching path. The BFS terminates when there are no unmatched odd-degree vertices left.

3.1   Results Comparison

Our experiments allow us to address two questions:

- is there a computationally-efficient (in terms of CPU time) solution to the CPP on large scale general graphs?
- how much distance can be saved using one-pass inspection strategy in comparison to a bi-directional approach?

The second question is of particular importance to the local authorities concerned.

To compare the computation time required to identify a CPT (inspection route) for each of the different graphs and variant approaches, the experiments were each running on a standard desktop PC: Intel i7-3770 CPU at 3.40 GHz with 24 GB memory under the Windows 7 operation system.

The exact methods using blossom algorithm in the matching process achieve the optimal CPP tour, whereas heuristic based methods can only find near-optimal solutions. Therefore, we analyse question two using the results generated

Algorithm 3. Greedy improvement algorithm, used to improve the matching result of the construction approach, algorithm 2 above. $l(m_i)$ is the distance between the two vertices in the matching m_i.

improved=true;
while improved **do**
 improve=false;
 for Every pair of matchings m_i, $m_j \in M$ **do**
  Generate new matchings m_k, m_l by exchanging vertices between m_i and m_j
  if $l(m_k) + l(m_l) < l(m_i) + l(m_j)$ **then**
   Replace m_i and m_j by m_k and m_l;
   improved = true;
   break;
  end if
 end for
end while
Return M

by the exact methods. Equation 5 is applied to normalise the difference and express it as a percentage distance saving:

$$saving = \frac{l(bi\text{-}directional) - l(one\text{-}pass)}{l(bi\text{-}directional)} * 100\% \tag{5}$$

On our chosen platform, the 12 experimental set-ups are run for each of the local authority road network graphs except Norfolk. The graph representation of the Norfolk road network is too large to process without our novel graph reduction step, and we have only applied the optimal solutions from the smaller networks to that for Norfolk.

CPU Time Comparisons. Figure 6 plots the CPU time taken. For each road network graph, the graph reduction and graph reconstruction times are negligible, and do not show up on this scale.

In all road networks, the methods applying graph reduction (m5 to m12) shows much lower overall CPU time, which demonstrates the importance of our graph reduction approach on graphs of this scale. After graph reduction, the exact methods can produce the CPT faster than any tested methods directly running on the original graph in all cases.

Table 3 gives numerical results for the blossom algorithm experiments on the three forms of graph for all seven local authority road networks (except for the un-reduced Norfolk graph). The numerical results again emphasise the reductions in CPU time due to graph reduction.

Some explanation is needed of an apparent anomaly in the Blackpool data. In the top left panel of Table 3, the number of edges (column m) is significantly higher for the fully-reduced graph (third row) than for the unreduced or partially reduced graphs. This arises from the removal of all even vertices

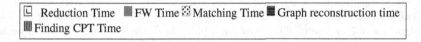

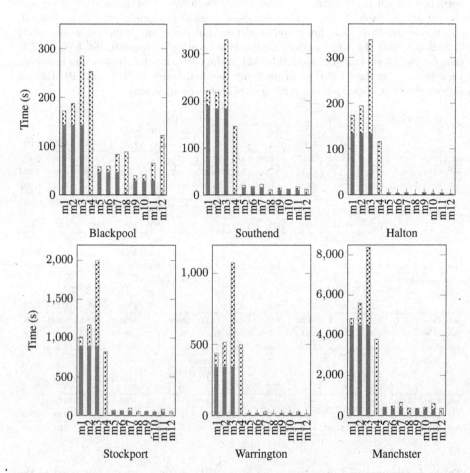

Fig. 6. Results of the 12 experiments for the six road networks. For each experiment, the bar shows the time taken, with shading showing the CPU usage of each algorithm (graph reduction and graph reconstruction take negligible time and are not visible in the plots).

in this network graph. In the Blackpool network, almost three percent of vertices have degree 4 or above, whereas no other network has more than 1.5% (Table 1). Whilst removal of degree-one and degree-two vertices always decreases graph complexity in terms of both the number of vertices and edges, removal of higher-degree vertices reintroduces significantly more edges. Fortunately, the CPU-hungry algorithms depend more strongly on the number of vertices.

Table 3. CPU time taken to calculate the CPT of each road network using the blossom algorithm for matching. The lower right panel gives the total CPU time for each road network for each experiment. The columns are as follows: n is the number of vertices (also shown in Table 1), and m the number of edges in the graphs after data cleaning, etc. RT is the CPU time for graph reduction: no reduction, reduction as described in Sect. 2.1, and the additional reduction in Algorithm 1, respectively. FW is CPU time for the Floyd-Warshall algorithm. MT is the CPU time for the blossom matching algorithm. CT is the CPU time to construct the final graph. CPT is the CPU time to extract the final inspection route using the Hierholzer algorithm.

	Blackpool (B)						Southend (SO)					
	n	m	CPU (s)				n	m	CPU (s)			
			RT	FW	MT	CT	CPT		RT	FW	MT	CT	CPT
m1	5103	7124	0	143.32	28.97	0.016	0.187	4571	5738	0	189.84	31.14	0.001	0.202
m5	3398	5419	0.14	45.54	11.11	0.016	0.234	2016	3162	0.14	16.05	4.26	0.001	0.203
m9	2772	10803	0.51	28.84	10.34	0.016	0.016	1746	5922	0.28	10.51	4.25	0.001	0.202

	Halton (H)						Stockport (ST)					
	n	m	CPU (s)				n	m	CPU (s)			
			RT	FW	MT	CT	CPT		RT	FW	MT	CT	CPT
m1	4796	5430	0	134.38	39.44	0.002	0.187	8665	10482	0	910.62	114.21	0.000	0.671
m5	1211	1852	0.15	2.96	1.97	0.001	0.218	3277	5063	0.45	55.91	12.46	0.006	0.826
m9	1148	1970	0.16	2.53	1.96	0.002	0.218	3004	5662	0.53	40.86	11.79	0.003	0.858

	Warrington (W)						Manchester (M)					
	n	m	CPU (s)				n	m	CPU (s)			
			RT	FW	MT	CT	CPT		RT	FW	MT	CT	CPT
m1	7471	8556	0	343.18	95.32	0.001	0.499	13337	16500	0	4438.04	355.83	0.013	1.58
m5	2136	3218	0.29	10.31	5.84	0.000	0.561	5842	8949	1.02	369.71	47.75	0.015	1.79
m9	2108	3265	0.31	9.78	5.83	0.001	0.561	5486	9622	1.12	306.29	46.71	0.015	1.67

	Norfolk (N)						Summary						
	n	m	CPU (s)				Total CPU(s)						
			RT	FW	MT	CT	CPT	B	SO	H	ST	W	M	N
m1	44912	53482	0	–	–	–	–	172	221	174	1025	439	4795	-
m5	15647	23904	18.27	4977.19	5499.49	0.171	34.523	57	21	5	70	17	420	10530
m9	14796	26487	19.66	4227.01	4759.01	0.182	34.881	49	15	5	54	16	356	9041

Table 4. Distance and distance saving of the CPT one-pass route, compared to the bi-directional route monitoring approach.

	Bi-directional (Euler Tour)	One-pass (CPT)	Saving		Average degree
			Distance	(Equation 5)	
Blackpool	1031 km	671 km	360 km	34.86%	2.79
Southend	1016 km	676 km	340 km	33.46%	2.51
Manchester	2631 km	1839 km	792 km	30.10%	2.47
Stockport	1891 km	1369 km	522 km	27.57%	2.42
Norfolk	26234 km	18268 km	7966 km	30.36%	2.39
Halton	1239 km	917 km	322 km	26.03%	2.26
Warrington	1758 km	1300 km	458 km	26.00%	2.29

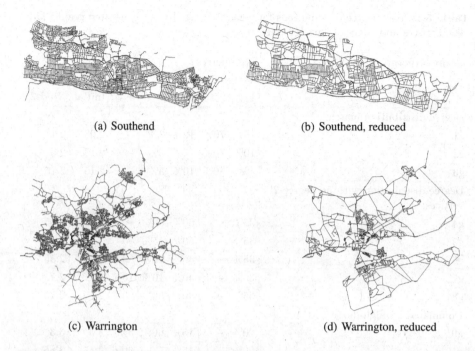

(a) Southend   (b) Southend, reduced

(c) Warrington   (d) Warrington, reduced

Fig. 7. Original maps and reduced graphs for Southend and Warrington road networks, illustrating grid-like and tree-like road networks.

Distance Comparisons. To compare totally distance vehicles should travel using one-pass versus bi-directional road inspection approach, Table 4 presents details of distance savings in these experiments. The distance saving for the one-pass inspection is between 26% and 35%.

Investigation of the differences in distance saved shows that, in addition to the added complexity of networks with high-degree vertices (noted above, and shown in the final column of Table 4), these reflect different road network topologies. Blackpool, Southend, Manchester and Norfolk, can be characterised as having predominantly grid-structured networks, which are conducive to efficient one-pass monitoring. By contrast, Warrington and Halton have predominantly tree-structured road networks which inevitably leads to visiting more streets twice, even in the one-pass case. To illustrate this, Fig. 7 gives the original and reduced-graph networks for Southend and Warrington.

4 Simulated Data Experiments

Our analyses of graphs representing real road network data lead to striking conclusions about the distance saving of one-pass, as compared to bi-directional, road inspection strategies. Although savings vary by about 10% points across different road networks, the distance savings were consistently above 26%.

Table 5. Vertex degree distributions for generated graphs. All generated graphs have 1000 vertices and edges of length 1 only.

Graph structures	Parameters						
	Vertex degree					Average degree	
	1	2	3	4	5	Initial	No d-2
Degree distribution match:							
g1	15%	70%	12%	3%	–	2.03	2.1
g2	10%	70%	15%	5%	–	2.15	2.5
g3	5%	80%	10%	5%	–	2.15	2.75
Degree distribution with no degree-2 vertices:							
g4	35%	–	60%	5%	–	2.35	2.35
g5	35%	–	50%	15%	–	2.45	2.45
g6	35%	–	40%	25%	–	2.55	2.55
g7	35%	–	40%	10%	15%	2.7	2.7
g8	35%	–	40%	5%	20%	2.75	2.75
Dominated by high-degree:							
g9	20%	–	15%	60%	5%	3.3	3.3
g10	20%	–	15%	5%	60%	3.85	3.85

To explore the interaction between graph layout and distance savings, we conducted a set of experiments on generated graphs using the blossom-based approach that was shown to be optimal above.

Random graphs were created using an algorithm proposed by [1]. Graphs are created with a fixed number of vertices (we choose 1000) to specified vertex-degree distributions. Our vertex-degree distribution parameters are shown in Table 5, which also summarises the average vertex degree of the generated graphs, before and after the initial removal of degree-two vertices. Three parameter settings of random graphs (g1 – g3) are generated with similar vertex-degree distribution to the graphs representing our real-world road networks. A further five parameter settings of random graphs (g4 – g8) have similar vertex-degree distribution to the graphs of real-world road networks after cleaning to remove degree-two vertices. There are also two parameter settings of random graphs that are dominated by degree-four (g9) and degree-five (g10) vertices. In the random generation, all edge lengths are set to one, and no attempt is made to generate graphs with particular structural characteristic (grid-like, tree-like). Our experiments focus on the generalisation of the influence of vertex degree, only.

For each group of graphs, we run these three sets of experiments with progressively greater reductions, dictated by our data cleaning and graph reduction approach to the road network graphs: 1) data cleaning to remove degree-two vertices, 2) graph reduction to remove all degree-one and degree-two vertices, and 3) further reduction to remove all even-degree vertices. For each experiment, we report the average of 30 runs, a number selected to give acceptable total run time but a suitably-low statistical error.

4.1 CPU Time Results

Table 6 presents details of the CPU time taken (using the computational platform described in Sect. 3.1) for the overall CPT identification process (vertex-pair distances using FW, blossom matching, graph reconstruction and CPT identification), and then for the three levels of data cleaning and graph reduction, on each of the 10 types of graph.

Each graph reduction makes a big contribution to CPU time reduction. The results for the first group of graphs, those generated to match the vertex degree distribution of the complete road network graphs, shows the importance of removing degree-two vertices in this respect. A further large time saving occurs when removing degree-one vertices – from Table 6 we can see that these make up 50% of the vertices in the example g1 graph after removal of the degree-two vertices. Comparing the improvement in CPU times between the full graph reduction (last column of Table 6) and the data-cleaning reduction of removing only degree two-vertices (middle column total), the results for this first group of graphs show CPU time savings of 95.5%, 90.3% and 78.5%, respectively.

Table 6. CPU time for CPT tour extraction on generated random graphs g1 – g10, showing the effect of graph reductions. Lowest overall CPU times for each graph type are in emboldened.

Graph structure	CPU(ms)						
	FW, blossom GR, HA	Removing degree 2		And removing degree 1		And removing all even degree	
		Reduction	Total	Reduction	Total	Reduction	Total
g1	19691.82	10.66	577.56	11.61	29.49	12.07	**25.91**
g2	19616.67	10.21	568.72	11.36	86.71	12.93	**54.86**
g3	19522.89	11.30	183.39	11.75	60.01	15.18	**39.37**
g4	20464.62	–	–	6.34	1831.37	7.76	**1436.3**
g5	20312.60	–	–	6.22	2086.82	14.64	**1062.11**
g6	20582.34	–	–	5.55	2528.90	863.95	**1666.97**
g7	20433.91	–	–	4.88	2780.43	13.07	**1414.97**
g8	20359.70	–	–	4.87	2810.04	10.99	**1622.76**
g9	20339.90	–	–	3.88	**8468.91**	248918.2	249369.4
g10	20407.10	–	–	3.68	9209.61	17.08	**4992.87**

The graphs in group g4 – g8 (and, indeed, g9 and g10) are generated without degree-two vertices. For these graphs, the graph reduction steps also show a very large CPU time reduction over the un-reduced time.

For the graphs dominated by higher-degree vertices, g9 and g10, the results again show that graph reduction leads to time reductions, but, for g9 which is dominated by degree-four vertices (60% of all vertices), CPT extraction with removal of all even vertices is almost three times the CPU time for just removing degree-one vertices, and somewhat greater than the CPU time for cleaned graphs with neither subsequent reduction step. In this case, the last column of Table 6 shows that there is a very large CPU time overhead for deleting the significant number of degree-four vertices and any other higher even-degree vertices. In contrast, the graphs dominated by degree-five vertices, g10, show a pattern that is consistent with the graphs that are similar to the real road network graphs.

Further analysis of specific graph results shows how these CPU time savings arise, since the FW and the blossom matching process are the most CPU-intensive parts of the CPT extraction.

Table 7. Degree distribution after each graph reduction for a typical g1 graph.

	Vertex degree				Total vertices
	1	2	3	4	1000
Original graph					
	15%	70%	12%	3%	
Delete degree 2	50%	–	40%	10%	300
Delete degree 1, 2	–	–	78%	22%	84
Delete degree 1, even	–	–	–	–	66

The complexity of FW is dependent on the number of vertices in the graph. Table 7 shows that degree-two removal (data cleaning) removes 70% of vertices, and subsequent graph reduction reduces 300 vertices successively to 84 then 66 vertices on which to run FW. It is the odd-degree vertices that influence the CPU time of blossom algorithm matching (step 2) – there are 270 odd-degree vertices in this example of a g1 graph, but we have the much smaller number, 66, of odd-degree vertices after removal of degree-one and all even-degree vertices. Thus we can conclude that the graph reduction contributes both to the reduced running time of the FW, shortest-path-between-pairs calculation, and to the reduced running time of matching process.

4.2 Distance Saving Results

Figure 8 shows the average distance savings (calculated using Eq. 5) and variances about the mean from our 30-run sets. As in the road network graphs, the greatest distance savings are associated with higher average degree graphs. The pattern also shows within groups of graphs with similar degree distribution (identified by shading in Fig. 8).

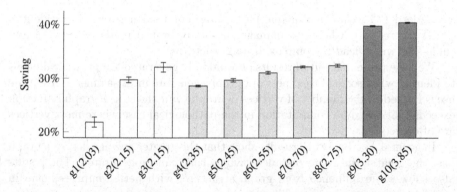

Fig. 8. Distance saving (Eq. 5 for the one-pass route compared to bi-directional routing on the randomly generated graphs g1 – g12. For graph characteristics, see 5. Numbers in brackets are the average vertex degree for that graph. Give 95% error bars.

These results on randomly generated graphs support our general observation of significant distance savings for a one-pass CPT over a bi-directional inspection tour, even though our generated graphs ignore the impact of different graph topologies and edge lengths. The only distance saving that is worse than the 26%-plus savings on the road network graphs is for g1, those graphs that have an average degree of close to two before and after removal of degree-two nodes. Comparing to other graphs, g1 graphs also have the highest proportions of degree-one vertices after data cleaning (removing degree-two vertices) – dead-ends that have to be traversed twice in order to continue a tour.

5   Conclusions

This paper explores the potential benefits for road inspection of using inspection vehicles that can collect both side of one road-lane information at the same time, as opposed to more traditional single-sided monitoring that has to use a bi-directional strategy to monitor every road.

By making various assumptions, we can systematically clean the map-based route data to replace omitted intersections. We create road network graphs from the seven UK Local Authority road network data sets, cleaning this data to remove degree-two vertices that simply record bends in the road. The most efficient bi-directional route is an Euler Tour of the original road network digraph with each 'edge replaced by two arcs in opposite directions. For the one-pass vehicles, we need to find the shortest route that traverses every edge of the undirected graph, a Chinese Postman Tour. We use the most efficient solutions available to extract a CPT, and explore heuristic approaches to the expensive time consuming step, the vertices matching.

A novel contribution of our work is to introduce graph reduction techniques. It is notably helpful on sparse graph like real-world road networks; even the graph of a UK county road network is amenable to CPT route calculation on a standard PC within reasonable CPU time. For road networks in residential areas, the reduction helps to manage the many branch roads, close road and culde-sacs which lead to complex graph structures.

We have presented initial results on randomly generated graphs that allow us to identify what sorts of road network graphs generate most savings – CPU time use is related to the number of vertices in graphs, but the cost of graph reduction exceeds that saved on computation in the pathological case where most vertices are of degree-four.

In general, road network results show that the greater average degree a graph has, the greater the route distance savings that can be generated. The results also show some influence from graph structure, with better improvements in grid-like structures than tree-like structures.

We present conclusive evidence that the one-pass approach offers significant savings over the bi-directional approach to road inspection. This assumes that the effects of data cleaning (which may wrongly insert new intersections, and may not be able to identify some missing road sections) are similar on both the one-pass and bi-directional tour calculations.

5.1 Estimated Cost Savings of One-Pass Road Monitoring

In co-operation with our business partner, Gaist Solutions Ltd., we can estimate the national savings of the one-pass strategy. Across all the results for the seven UK Local Authority road networks, we find a distance-saving of roughly 30%. If we assume that the seven case studies are typical of the UK, then we can use this to estimate the total annual saving for UK local authority spending on road inspection using a one-pass strategy. We use published data, rounded to avoid a false appearance of precision in the estimates.

The whole UK road network under local authority control (i.e. without trunk roads and motorways) is roughly 240,000 km [25], broken down to 30,000 km of major roads and 210,000 km of minor roads. The UK highway management authority states that road inspection should cover all major and about one-third of minor roads each year [4].

Extrapolating from our seven examples, we can estimate that a bi-directional strategy requires annual inspection route distances totalling roughly 60,000 km of major roads (twice the total distance) and 140,000 km minor roads (twice the total distance of one-third of the routes each year).

If a one-pass strategy results in a 30% distance saving, then the inspection route distance reduction would be roughly 18,000 km for major roads and 42,000 km for minor roads. According to the cost information provided by our industry partner, Gaist, we can estimate a cost saving using a linear function that includes labour, petrol, equipment and vehicle maintenance costs. This gives an approximate cost of £30 per km of major road and £90 per km of minor road. The cost is higher for minor roads because running speed is lower, and there are more likely to be obstructions and blockages. We can therefore tentatively suggest an annual saving across all UK Local Authorities of $30 * 18000 + 90 * 42000 = £4.32$ million.

At this stage, we have an overview of how much cost saving could be made by using a one-pass inspection strategy instead of a bi-directional strategy. Gaist is now using our approach, including graph reduction in planning its national scale road inspection programme. However, there are more practical issues to address, such as the number of one-pass inspection vehicles required to inspect the network, and identification of optimal starting points for each vehicle tour.

Acknowledgements. The authors would like to thank Gaist Solutions Ltd. for providing data and domain knowledge. This work has been funded by the Large Scale Complex IT Systems (LSCITS) EngD EPSRC initiative in the Department of Computer Science at the University of York.

References

1. Blitzstein, J., Diaconis, P.: A sequential importance sampling algorithm for generating random graphs with prescribed degrees. Internet Math. **6**(4), 489–522 (2011)
2. Chartrand, G., Oellermann, O.: Graph minors. In: Applied and Algorithmic Graph Theory, pp. 277–281. McGraw-Hill (1993)
3. Christofides, N.: Worst-case analysis of a new heuristic for the travelling salesman problem. Carnegie-Mellon Univ Pittsburgh Pa Management Sciences Research Group (1976)
4. Department for Transport: Well-maintained Highways: Code of Practice for Highway Maintenance Management (2005)
5. Dijkstra, E.W.: A note on two problems in connexion with graphs. Numer. Math. **1**(1), 269–271 (1959)
6. Edmonds, J.: Paths, trees, and flowers. Can. J. Math. **17**(3), 449–467 (1965). http://www.cs.princeton.edu/introcs/papers/edmonds.pdf
7. Edmonds, J., Johnson, E.L.: Matching, euler tours and the chinese postman. Math. Program. **5**(1), 88–124 (1973). http://link.springer.com/article/10.1007/BF01580113
8. Even, S.: Paths in graphs. In: Even, G. (ed.) Graph Algorithms, pp. 1–28. Cambridge University Press, Cambridge (2011). http://xueshu.baidu.com/s?wd=paperuri%3A%287ed99be525cb1711f144757e32844ef9%29&filter=sc_long_sign&tn=SE_xueshusource_2kduw22v&sc_vurl=http%3A%2F%2Fdl.acm.org%2Fcitation.cfm%3Fid%3D2049721&ie=utf-8&sc_us=9448669111505408634
9. Fleury, M.: Deux problemes de geometrie de situation. J. Math. Elementaires **2**(2), 257–261 (1883)
10. Floyd, R.W.: Algorithm 97: shortest path. Commun. ACM **5**(6), 345 (1962)
11. Fredman, M.L., Tarjan, R.E.: Fibonacci heaps and their uses in improved network optimization algorithms. J. ACM **34**(3), 596–615 (1987). http://dl.acm.org/citation.cfm?id=28874
12. Gabow, H.N.: Data structures for weighted matching and nearest common ancestors with linking. In: Proceedings of the First Annual ACM-SIAM Symposium on Discrete Algorithms, pp. 434–443. Society for Industrial and Applied Mathematics (1990). http://dl.acm.org/citation.cfm?id=320229
13. Guan, M.: Graphic programming using odd or even points. Chin. Math. **1**(110), 273–277 (1962)

14. Hierholzer, C., Wiener, C.: Ueber die Möglichkeit, einen Linienzug ohne Wiederholung und ohne Unterbrechung zu umfahren. Math. Ann. **6**(1), 30–32 (1873). http://dx.doi.org/10.1007/BF01442866, http://www.springerlink.com/index/X4458623778T4704.pdf

15. Irnich, S.: Solution of real-world postman problems. Eur. J. Oper. Res. **190**(1), 52–67 (2008). http://linkinghub.elsevier.com/retrieve/pii/S0377221707005486

16. Jha, M.K., Udenta, F., Chacha, S., Abdullah, J.: Formulation and solution algorithms for highway infrastructure maintenance optimisation with work-shift and overtime limit constraints. Procedia Soc. Behav. Sci. **2**(3), 6323–6331 (2010). http://linkinghub.elsevier.com/retrieve/pii/S1877042810010943

17. Jha, M.K., Udenta, F., Chacha, S., Karri, G.: A modified arc routing problem for highway feature inspection considering work-shift and overtime limit constraints. New Aspects of Urban Planning and Transportation, pp. 105–109 (2008), http://w3.ualg.pt/~tpanago/public/urban-planning-and-transportation.pdf#page=108

18. Kallen, M., Van Noortwijk, J.: Optimal periodic inspection of a deterioration process with sequential condition states. Int. J. Press. Vessels Pip. **83**(4), 249–255 (2006). http://www.sciencedirect.com/science/article/pii/S0308016106000251

19. Kolmogorov, V.: Blossom V: a new implementation of a minimum cost perfect matching algorithm. Math. Program. Comput. **1**(1), 43–67 (2009). http://link.springer.com/10.1007/s12532-009-0002-8

20. Kurtzberg, J.M.: On approximation methods for the assignment problem. J. ACM (JACM) **9**(4), 419–439 (1962). http://dl.acm.org/citation.cfm?id=321140

21. Laporte, G.: Modeling and solving several classes of arc routing problems as traveling salesman problems. Comput. Oper. Res. **24**(11), 1057–1061 (1997). http://www.sciencedirect.com/science/article/pii/S0305054897000130

22. Lovász, L.: Graph minor theory. Bull. Am. Math. Soc. **43**(1), 75–86 (2006). http://www.ams.org/bull/2006-43-01/S0273-0979-05-01088-8/

23. Madanat, S., Ben-Akiva, M.: Optimal inspection and repair policies for infrastructure facilities. Transp. Sci. **28**(1), 55–62 (1994). http://pubsonline.informs.org/doi/abs/10.1287/trsc.28.1.55

24. Maji, A., Jha, M.: Modeling highway infrastructure maintenance schedules with budget constraints. Transp. Res. Rec. J. Transp. Res. Board **1991**(1), 19–26 (2007). http://trb.metapress.com/index/6U27625743810286.pdf

25. Murphy, A.: Road Lengths in Great Britain: 2013 (2014)

26. Reingold, E.M., Tarjan, R.E.: On a greedy heuristic for complete matching. SIAM J. Comput. **10**(4), 676–681 (1981). http://epubs.siam.org/doi/abs/10.1137/0210050

27. Smilowitz, K., Madanat, S.: Optimal inspection and maintenance policies for infrastructure networks. Comput. Aided Civ. Infrastruct. Eng. **15**(1), 5–13 (2000). http://onlinelibrary.wiley.com/doi/10.1111/0885-9507.00166/abstract

28. Solomonik, E., Buluç, A., Demmel, J.: Minimizing communication in all-pairs shortest paths. In: 2013 IEEE 27th International Symposium on Parallel & Distributed Processing (IPDPS), pp. 548–559. IEEE (2013). http://ieeexplore.ieee.org/xpls/abs_all.jsp?arnumber=6569841

29. Thimbleby, H.: The directed chinese postman problem. Softw. Pract. Experience **33**(11), 1081–1096 (2003). http://doi.wiley.com/10.1002/spe.540

Assessment of Risks in Manufacturing Using Discrete-Event Simulation

Renaud De Landtsheer[1], Gustavo Ospina[1], Philippe Massonet[1],
Christophe Ponsard[1(✉)], Stephan Printz[2], Sabina Jeschke[2], Lasse Härtel[3],
Johann Philipp von Cube[3], and Robert Schmitt[3]

[1] CETIC Research Centre, Charleroi, Belgium
{rdl,go,phm,cp}@cetic.be
[2] Institute for Management Cybernetics (IfU),
RWTH Aachen University, Aachen, Germany
{stephan.printz,sabina.jeschke}@ifu.rwth-aachen.de
[3] Fraunhofer Institute for Production Technology (IPT), Aachen, Germany
{lasse.haertel,philipp.von.cube,robert.schmitt}@ipt.fraunhofer.de

Abstract. Due to globalisation, supply chains face an increasing number of risks that impact the procurement process. Even though there are tools that help companies address these risks, most companies, even larger ones, still have problems for adequately quantifying the risks on their current process as well as on alternative process. The aim of our work is to provide companies with a software supported method for quantifying procurement risks and establishing adequate strategies for risk mitigation at an optimal cost. Based on the results of a survey on risk management practices and industrial needs, we developed a tool that enables them quantifying these risks. The tool makes it easier to express key risks via a process model that offers an adequate granularity for expressing them. A simulator incorporated in our tool can efficiently evaluate these risks through Monte-Carlo simulation technique. Our main technical contribution lies in the development of an efficient Discrete Event Simulation (DES) engine, together with a Query Language that can be used to measure business risks from the simulation results. We show the expressiveness and performance of our approach by benchmarking it on a set of cases that are taken from industry and cover a large set of risk categories.

Keywords: Discrete Event Simulation · Manufacturing · Supply chain · Procurement risks · Risk management

1 Introduction

Companies are faced with increasing procurement risks within the context of a global economy. These risks can be related to many different factors, such as, for example, the geographic location and the political and economic situation of the involved parties (suppliers, warehouses and factories). Assessing these risks

© Springer International Publishing AG 2017
B. Vitoriano and G.H. Parlier (Eds.): ICORES 2016, CCIS 695, pp. 201–222, 2017.
DOI: 10.1007/978-3-319-53982-9_12

alone is a difficult task as the risks may only reveal themselves at the end of the production chain and it also requires consideration be given to the impact of internal risks, as well as thought be given to the complexity of the manufacturing process (which could decrease the capacity for adaptation in the case of supply failure) or the level of optimisation in place (which could increase the impact in case of disruption).

Helping company managers make the right decision in the face of risks is not an easy task. Whilst analytical reasoning is impractical, model-based simulation has proved to be a competent approach [1]. Procurement risks present extra challenges as they occur at one end of the process (output), but can sometimes only be measured at the other end (input), therefore they require thought throughout the whole manufacturing process. The scope of our work is to address this challenge, focusing on small and medium enterprises in the field of mechanical engineering.

Our ultimate goal is to produce a user-friendly, tooled methodology that will guide the user through the whole process of risk assessment. In order to reach this goal, our work is structured as follows:

- A taxonomy of supplier and internal risks is identified, starting with the simplest risk - a shortage of raw materials, which can eventually lead the whole process chain to more elaborate risks, depending on the kind of order policy used.
- A survey was conducted on the practice of risk evaluation in an industrial context [2]. The results of this survey showed that nearly 66 % of the companies perform risk evaluations, although only 10 % rely on dedicated software tooling. This means that in practice risks are evaluated by an individual estimation of the cost factor and the probability of occurrence. In general, and by including historical data in the estimation, the quality of the estimation improves. However, relying on historical data and estimating the impact of factors, like delivery times, means that material quality is impossible to estimate, even in the case of changing suppliers or adding parallel processes to the chain. Based on the requirements identified in the survey, a software based risk management framework was defined.
- We developed a modelling and simulation toolset for identifying risks, quantifying them and deciding on design alternatives that can help mitigate risks. The main technical scope of this paper is to detail our toolset framework and show how it helps focus be retained on the risks during modelling, so as to stay efficient during the modelling time, simulation time and result analysis time.
- Finally, we are validating our work through a group of companies that are already trying out our tool via an easy to use web interface. Although this validation is not yet complete, we can already benchmark our approach on a number of industrial cases and in doing so assess the expressiveness and performance of our approach.

Our modelling framework includes concepts such as *storages*, where items can be stored or retrieved with a maximum capacity, and several types of production *processes*, each with different timing and failure behaviours. In addition, we defined a query framework on models that are fully declarative and include arithmetic, temporal and logic operators, as well as basic probes for the elements of our factory model (contents of a storage, whether a process is running or not, etc.). Based on this query language, the software tool is able to calculate the probabilities of different scenarios (e.g. delay in deliveries, defective parts or poor quality) and their impact, which is all based on a timed model of the relevant factory process.

This approach to monetary risk quantification is based on an approach developed in the Q-Risk project [3]. The simulation toolkit relies on the discrete event simulation module of the OscaR framework, particularly its simulation layer, and adds dedicated abstractions that are dedicated to the timed modelling of factories, and the modelling of risk-related queries [4].

Our main contribution lies not only in the risk-driven dimension of our framework, but also in regards to the usability factors. Its design is based on a number of trade-offs between the expressiveness and simplicity of the modelling language, as well as efficiency of the simulation engine.

The paper is structured as follows: Sect. 2 presents the context of our work; Sect. 3 presents our modelling language for representing factories; Sect. 4 presents our query language that can serve to evaluate risks; Sect. 5 illustrates how complex risks can be included in our query language; Sect. 6 describes our DES prototype and more specifically the Query Engine. Section 7 shows the benchmarking of our simulation tool both on the expressiveness and performance dimensions; Sect. 8 discusses some related work; Sect. 9 concludes the paper.

2 Background

In order to assess and quantify different kind of risks in manufacturing processes, we model the manufacturing process as a flow graph. This model captures the key procurement steps and the production process itself. Resource storage (in warehouses or stockrooms) and the flow of raw materials in basic processes will be explained in Sect. 3.1. The main graphical notations implemented by the graphical part of framework are shown in Fig. 1, which is a model used later in the benchmarking process. Notations are quite self explanatory: a supplier is a little truck, storage types are represented by different variations of cylinders (the one with vertical bars can overflow) and processes are depicted with the industry icon (also with some variants: multiple horizontal lines means parallel batches, the cross means possible failure, the rounded, the rounded box depict a conveyor belt).

The operation of the whole manufacturing process can be described as a sequence of timed events. For instance, the first event involved with simulating a factory is fetching some materials for storage. This action can trigger a new order being sent to a supplier if the storage level reaches a certain threshold - in accordance with supply chain policy.

In the rest of this section we recall the nature of risks and the goal of risk management, then we give some details about Discrete Event Simulation and

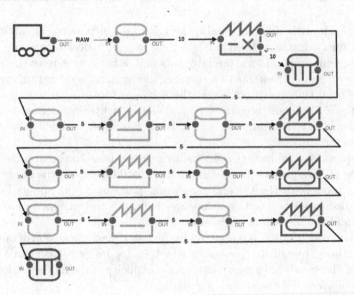

Fig. 1. Beer game model.

why it provides an adequate framework for modelling the operations of manufacturing processes and quantifying risks [5].

2.1 Risks and Risk Management

In order to develop a Discrete Event Simulation approach for quantifying the impact of risks in manufacturing enterprises, the nature of risk and the underlying process of risk management needs to be understood in detail.

Risk. Risks strongly affect an enterprise's business success and are directly related to costs, effort and yield [6]. Thereby, risk is understood as an event likely to occur with an undesired consequence. The most common and, for the approach, most convenient categories of risks are the cause and impact-oriented type. The root-cause-oriented approach considers information uncertainty and validity as risk [7]. The chance of not meeting a planned target is understood as an impact-oriented risk. However, only combining both categories of risk leads to the necessary scope of information needed to properly manage risks. Hence, risks need to be understood as having a certain likelihood of missing a defined target. Hence, the concept of risk is defined through three components: the *hazards*, or potential dangers, the *consequences* of those hazards, and their predicted frequency, or *likelihood* [8]. A "natural" quantification of the hazards associated with a risk is the quantification of product consequences in order of likelihood. A cybernetic model of procurement-based hazards and their management is presented in [9].

Risk likelihood can be modelled with probability distributions [10], as the occurrence of a risk hazard in a process or system is *uncertain*. In [11], a theory of probabilistic risk analysis is developed, which is associated with the concept of system reliability. As a risk is defined as the deviation from a planned value, statistical measures can thus be applied to operationalise and compare the possible magnitude of such deviations [12]. Evaluation of the risk analysis and the reliability of a system can be done with the Monte-Carlo method [1].

Risk Management. The main objective of risk management lies in the assessment of major corporate goals in regards to risk policy strategies. Hence, risks affecting long lasting business success need to be controlled. However, enterprises will never be able to totally eliminate risks and will always have to consider a certain degree of residual risk [13]. One key task of risk management is to identify and analyse risks as early as possible in order to take cost optimal risk treating actions [6].

The basic process of risk management (Fig. 2) is described in standards ISO 31000 and ONR 49000 ff. IEC 31010 provides an overview of corresponding risk management methods and techniques for a specific process.

2.2 Discrete Event Simulation

There are two main approaches for representing time if we want to simulate the behaviour of a system: the first approach is to use a *continuous time*, in which the events affecting the system occur as time "ticks", all of which is proportional to the actual expected time of system operation. The other approach is to have a *discrete time*, and concentrate the simulation on only the operational events, rather than the time. This is the basis of Discrete Event Systems (DES).

In the literature [14,15], the main components of a DES model are described as: *entities* (which are the items that are flowing and transformed throughout the

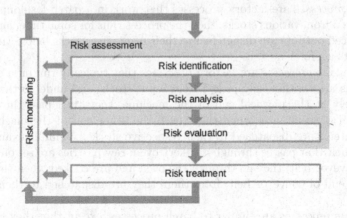

Fig. 2. Risk management process.

simulation), *queues* (representing storage devices or other areas in which entities wait to be used), *activities* (that actually perform some work on the entities), and *resources* (a special kind of entity that is required to operate activities).

DES models define *events* as discrete points in time in which the system state changes. The simulation of the model is simply a queue of different events triggered by the previous event, in other words the "next-in-time". Checking an event can trigger other events in the queue. For instance, checking the event starting an activity will trigger the events of fetching the corresponding entities needed to perform the activity, and thus ending the activity. The event of activity failure can also be triggered by a given probability.

Several software solutions exist to support DES based modelling for a variety of applications. Among the commercial software, we can cite AnyLogic [16], Arena [17] and Plant Simulation [18].

3   A Simulation Meta-Model for Factories

All the main elements of manufacturing processes are represented in our simulation meta-model, which allows us to define concrete models that are simulated in a Discrete Event Simulation engine. In addition to this, we designed a Query Language over concrete simulations in order to collect and analyse data.

3.1   Modeling Factory Processes

This section introduces the basic blocks for representing factories. In our approach, factories are modelled as a flow of items, processes and stocks.

Storages represent any kind of stock device or place for raw materials, like a warehouse, a barrel, a silo or a dumpster. They have a maximum capacity. When this capacity is reached, they either overflow, or create a blockage in the process, all of which is dependent on the settings of the storage. If a full stock storage overflows, any unloading material of that stock is lost.

Batch processes are factory processes that work in a batch fashion; supplies are collected from various stocks, then the process runs for some time, and finally the produced outputs are dispatched to their respective stocks before this whole cycle starts again.

Continuous processes are factory processes that typically run on a conveyor belt. Items are continuously picked from input stocks and undergo the process immediately at the physical end of the machine, they then pass through the machine in a queue, and when they reach the other end of the machine, the resulting items are dispatched to their respective stocks. A simple example is a conveyor belt that passes through a bakery oven; raw pastries are set on one end of the conveyor belt; they go through the oven and are cooked when they reach the other end of conveyor belt, from there they are dispatched to their output storage.

Splitting processes are similar to batch processes, except that they have several sets of outputs and when completed, one set of outputs is selected and the produced items are dispatched to the stocks associated with the selected output. This represents a quality assurance process, whose item flow is split into two (or more) separate flows based on the result of the quality assurance analysis.

Parallel processes are variants of the processes above, where several lines of the same process are running in parallel. Basically, all processes introduced here have a parameter specifying the number of process lines running in parallel.

Items flowing in processes and stocks are indistinguishable at any given point of the factory process as they all share the same part number. Yet, they have some intrinsic features: some items might come from a given process, others might be made out of poor quality supplies, etc. These intrinsic features can influence the behavior of some processes, such as the splitting process representing a quality assurance process. This notion of intrinsic features leads us to distinguish between two different types of storage, namely: First In-First Out (FIFO) storage and Last In-First Out (LIFO) storage.

3.2   Process Activation and Supply Chain Policies

Supply chain policies are also integrated in our model of the factory, together with activation policies that are able to turn a process on or off, depending on the demand for the output stock. To model these two concepts, we introduce the notion of *activable* and *activation*. An activable is something that can be enabled through an activation. We also associate a magnitude with the activation, or in other words, an integer. An activable can be a process or a supply order. In the case of a process, the activation represents the number of batches that the process is allowed to execute. In the case of an order, the magnitude represents the number of ordered items.

In our model, an order is a stationary, activable object that represents a class of order that can be passed. The order is passed when the modelled order object is activated.

Activables can be activated based on various rules that are also part of our modelling framework. There are three types of activation rules, namely: regular activations (performs the activation on a regular basis over a period of time);

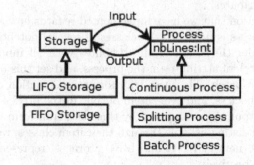

Fig. 3. Concepts of our process modelling languages.

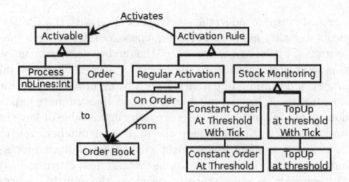

Fig. 4. Concept for modelling activation rules.

order-based activations (performs the activation when an order is received); and stock monitoring activation (performs the activation when the stock level gets below the threshold).

3.3   Modelling Intrinsic Item Features

To represent these intrinsic features, we introduce the notion of *item class*, representing the set of intrinsically identical items. Item classes are characterised by a set of boolean *attributes*. A global set of attributes is defined on the whole simulation model; each item has its own attributes, which is a subset of the global set of attributes. This set of attribute attached to the item defines the item class to which it belongs.

When an item flows through a process, the process can update the attribute of the item to reflect the process that was applied to this item. Similarly, when an item flows through a splitting process, the selected output can be specified according to the attributes of the item. At this point, we had to set a trade-off between the expressiveness of the modelling language, its simplicity, and the efficiency of the simulation. Processes can update the class of items through three basic operations: setting an attribute, clearing an attribute, or loading a constant set of attributes.

Another restriction that we have implemented regards how the class of items produced by a process is linked to the classes of several potential inputs of this process. We consider the union of all attributes with all inputs performed in order to start a batch of attributes in the process, and set this union as the start class for the whole batch. The class transformation function of the process is then applied to this class, and every item output in the process from this batch share the same output class computed by this class transform function.

Since items are distinguishable through their item classes, we introduced two models of storages, namely FIFO and LIFO storages, representing queues and stacks of items, respectively.

4   Performing Queries over Simulations

The goal of our approach is to perform risk-related queries on factory simulations. These queries are meant to be performed on single runs of simulations occurring inside the Monte-Carlo engine, which aggregates the query results over the runs. It can then be queried afterwards e.g. for mean, median, extremes, variance of these queries over the runs.

Our query language can roughly be split into three sets of operations, namely: probes on processes, probes on storages, and operators. Operators are split into five sets, namely logic operators, temporal logic operators, arithmetic operators, temporal arithmetic operators, and history recording operators. Arithmetic and logic operators differ by their their return types; they return numeric and boolean values respectively.

Since this query language runs over simulated time, we assume that the value of the queries are computed at the end of the trace on which they are evaluated. We define the operator of our language together with their semantics by using the $\models$ notation: $t \models P$ is the value of expression P when evaluated at position t of the current trace.

Some temporal operators refer to the previous position in time, denoted as $\mathrm{prev}(t)$, notably to compute deltas or assess changes. These should be used with care since we are in an event-based model of time, so adding such operators in the query will add extra time events in the simulation.

4.1   Probes for Processes

The probes on processes are atomic operators that extract basic metrics from processes of the simulation model. Suppose that p is such a process, the following probes are supported:

- $t \models \mathrm{running}(p)$ true if the process is running at time t, false otherwise.
- $t \models \mathrm{completedBatchCount}(p)$ the total number of batches performed by the process between the beginning of the trace, and time t.
- $t \models \mathrm{startedBatchCount}(p)$ the number of batches started by the process between the beginning of the trace, and time t. For a process with multiple lines, it sums up the started batches of each line.
- $t \models \mathrm{totalWaitDuration}(p)$ the total duration where the process was not running between the start of the trace, and time t. for a process with multiple lines, it sums up the waiting time of each line.
- $t \models \mathrm{anyBatchStarted}(p)$ true if a batch as started by the process at time t.

4.2   Probes for Storages

The probes on storages are atomic operators that extract basic metrics from storages of the simulation model. Suppose that s is such a storage:

- $t \models$ empty(s) true if the storage s is empty at time t, false otherwise.
- $t \models$ content(s) the number of items in the storage s at time t.
- $t \models$ capacity(s) the maximal capacity of s. This is invariant in time.
- $t \models$ relativeCapacity(s) the relative content of storage s at time t, that is: the content of the stock divided by the capacity of the storage.
- $t \models$ totalPut(s) the number of items that have been put into s between the beginning of the simulation and time t, not counting the initial ones.
- $t \models$ totalFetch(s) the number of items that have been fetched from s between the beginning of the simulation and time t.
- $t \models$ totalLostByOverflow(s) the number of items that have been lost by overflow from s between the beginning of the trace, and time t. If s is a blocking storage, this number will always be zero.

4.3 Operators

Logical Operators

- $t \models$ true the constant true.
- $t \models$ false the constant false.
- $t \models !l$ the negation operator.
- $t \models l_1$ op l_2 where *op* is one of $\{\&, \|\}$ represent conjunction, and disjunction operators, respectively, returning their conventional results.
- $t \models a_1$ comp a_2 where *comp* is one of $\{<, >, \leq, \geq, =, \neq\}$ represent comparison operators over numerical values, returning their standard results.

Temporal Logic Operators

- $t \models$ hasAlwaysBeen l true if for each t' in $[0; t]$, $t' \models l$
- $t \models$ hasBeen l true if there is a t' in $[0; t]$ such that $t' \models l$
- $t \models l_1$ since l_2 true if there is a position t' in $[0; t]$ such that $t' \models l_2$ and for each position t in $[t, t]$, $t \models l_1$
- $t \models @l$ true if both $t \models l$ and prev$(t) \models !l$.
- $t \models$ changed(e) e might be a logic or arithmetic expression; this evaluate to true when $t \models e$ and prev$(t) \models e$ have different values.

Arithmetic Operators

- $t \models n$ where n is a numerical literal represents a literal constant value
- $t \models a_1$ op a_2 where *op* is one of $\{+, -, *, /\}$ represent the classical arithmetic operators over numerical values, returning their conventional results.
- $t \models -a$ represents the unary negation.

Temporal Arithmetic Operators

- $t \models$ delta$(a1)$ is a shorthand for $t \models a$ prev$(t) \models a$
- $t \models$ cumulatedDuration(b) let be $T = (t_1, t_2) \| t_1 = prev(t_2) \& t_1 \models b \& t_2 \models b$ the accumulated duration of b is the sum over the couples (t_1, t_2) in T of t_2, t_1

- $t \models$ time evaluates to t.
- $t \models \min(a)$ the minimum over all the values of $t' \models a$ with t' in in $[0; t]$
- $t \models \max(a)$ the maximum over all the values of $t' \models a$ with t' in in $[0; t]$
- $t \models \mathrm{avg}(a)$ the average of all the values of $t' \models a$ with t' in in $[0; t]$
- $t \models \mathrm{integral}(a)$ the integral of $t' \models a \; dt'$ with t' in $[0; t]$. The integral is computed through the trapezoidal rule taking the events as discretisation base.

History Recording. Two operators are available for recording the evolution of a query throughout the simulation run. The type of these operators is a history of value. The value itself is Boolean, or arithmetic, respectively. These operators cannot be composed with any other operator. Semantically, they are only evaluated at the end of the run, so that there is no position in time in their definition.

- $\mathrm{record}(a)$ is the history of the value of the arithmetic query a over the simulation run.
- $\mathrm{record}(b)$ is the history of the value of the boolean query b over the simulation run.

5   Modelling Risks as Queries

Our query language allows for computation of any metric measurement given in the simulated factory model, and metrics associated to risks in particular. Thus, it is necessary to identify the risks we want to assess. The risk identification process is partly generic because there are well-known categories of risks. We considered here the classical delay/quality/quantity triangle, which can be defined as follows:

- **Delay risks** are related to the possibility of taking more time than expected to produce goods, for instance due to process malfunction, or problems in stocking materials due to supply and storage failures.
- **Quantity risks** are related to the possibility of producing less goods than expected in a process or losing materials in the storage stage.
- **Quality risks** are related to the possibility of a process producing "bad" goods, or the degradation of materials whilst in storage.

The characteristics of these risks, such as their likelihood and their impact, may vary a lot. We quantify these risks by expressing them in our query language, generally as a cost value. Risks can occur at different levels of granularity. *Elementary risks* occur at component level (such as a provider, a storage type, a process) but not necessarily result in a system level risk if the supply chain is designed to cope with the risk. *System-level risks* occur at system level, for example, the global production latency, or the critical path of a supply chain.

In the rest of this section, we present how typical risks can be expressed in our query language. Queries can reference other queries; this makes it possible to refer fine grained risks in the definition of higher level risks.

Risks Specific to Suppliers. For a single supplier, the risk model is directly encoded in terms of quality, quantity and delay, so it makes little sense to measure it. We can consider more complex configurations, such as a combination of suppliers, reactive suppliers with less quality or quantity, and a slower supplier with high quality or quantity. We can assess the adequacy of the design of a supplier combination by measuring the probability of underflow in the common input storage they are supposed to replenish, e.g. using the probe cumulatedDuration(content(sto) > MIN), where MIN is some minimal "safety" stock.

Risks Specific to Storages. As a delay risk for storage elements, we can compute the amount of time it takes before a storage overflow occurs ovst with the probe cumulatedDuration(hasAlwaysBeen(ovst) < 1). This allows us to know whether the storage has been losing pieces for a long time and be able to estimate the extent to which the storage should be enlarged.

The most basic quantity risk for an overflowing storage is the number of lost pieces. This is measured by the probe totalLostByOverflow(ovst). Together with the previous probe, useful adaptations to the size of the storage container can be decided.

With respect to the quality risks, it is possible to find that a storage is oversized, that is that the maximum contents of the stock throughout the simulation are a lot lower than the storage's capacity. This can be measured with the probe max(relativeCapacity(stock)) and it can be used to verify whether that value is higher than an acceptable percentage number.

Risks Specific to Processes. For quantity risks, we are interested in processes that do not work enough in the simulation. The percentage of idle time for a process p is measured by the probe (totalWaitDuration(p)/currentTime) * 100. To detect whether the process p operated at all in the simulation, we can use the probe: hasAlwaysBeen(!anyBatchStarted(p)).

About quality risks, a basic query is used to look at the ratio of failed part from a failing batch process: totalPut(p)/completedBatchCount(p). This is however of limited interest as it will return a value close to the encoded probability of failure. More general attributes are however being defined on processes with more complex transformation functions - this includes more randomness, which would make such a query informative at that level.

System Level Risks. Quantifying risks at system level is highly dependent of each considered case. Let us illustrate some typical queries that can be used. A general consideration is that at system level risk are systematically turned into costs so they can be compared. In order to do this specific attributes are available: for example the cost of all parts going through stock, the cost to repair a damage machine, penalties for supplier being late, etc.

For quality risks, considering losses occur from failing processes FP1 and overflowing storage ST1. Considering there is no rework, the query LOSS is:

totalFailed(FP1) * partCost(FP1)+totalLostByOverflow(ST1) * partCost(ST).
Assuming there is also a similar query to define the VALUE_ADDED (based
on the cost of incoming parts versus produced parts), the profitability is then
defined as VALUE_ADDED/(LOSS + VALUE_ADDED). These queries could be
made more general using a component selector to select all relevant parts to use
during specified computation.

For quantity risks, an OUTPUT storage can be compared with an ORDER
book: (content(OUTPUT) − content(ORDER)) * partCost(ORDER). We use
the agreed order cost which may differ from the cost of the produced parts.

Global delay risks are more difficult to express because the DES engine does
simulate the flow of parts but only reports transformation events. It is however
possible to monitor the time required to fulfill some order and compare it with
the agreed delivery time and take into account possible penalties of being late.

6   Implementation

In this section, we first give an overview of the global architecture of the software
tool before giving more details about the execution of queries in the simulation,
which is the main body of our work.

6.1   Global Simulator Architecture

The simulator is implemented using the OscaR DES module [4] and is written
in Scala [19]. A modelling web front-end was developed with JavaScript technologies, mainly Bootstrap [20], JQuery [21] and JointJS [22]. The lightweight
Scalatra [23] web framework was used to wrap up the simulator as a set of web
services. The web server contains a Monte-Carlo simulation engine that is able
to aggregate the results (especially the queries) of several simulations over the
same model. Figure 5 illustrates the architecture.

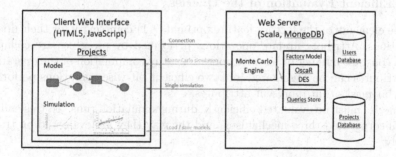

Fig. 5. Global architecture of the software tool.

6.2 Supporting Risk Identification

The software tool includes a wizard helping the user to express risk using our query language. A succession of sreenshot of this wizard is shown in Fig. 6. It starts from the main risk categories (quantity/quality/delay) then guides the user into specifying how the risk can be measured both at system level and then at component level. A number of predefined queries are available in each context. Queries can also be edited and designed from scratch using a plain text editor. The figure shows the encoding of a quantity risk measuring the number of lost pieces in a factory.

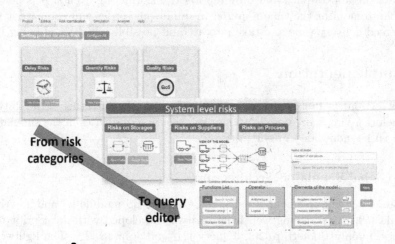

Fig. 6. Encoding a query for a quantity risk.

6.3 Efficient Evaluation of the Queries

All the elements of the factory feature optimal $O(1)$ complexity for their update operations. Attribute update operations performed by processes are collapsed into a constant number of bit-wise operation. Any combination of the operations on attributes can be aggregated into two efficient bit-wise operations performed using bit masks that represent attributes.

Queries must be evaluated efficiently during simulation runs. To this end, we have incorporated three mechanisms and the way they are evaluated on traces, namely:

– Incremental evaluation throughout the simulation run.
– Minimal updates that allows updating only the relevant fragment of queries.
– Bottom-up updates to allow the sharing of sub-queries.

Incremental Evaluation. Queries are evaluated incrementally throughout the simulation, so that no simulation trace is actually generated. At each step of the simulation, queries are notified about the new step, and they update their value accordingly.

A very simple expression illustrating this is `a+b`. At each step t, the value of this expression is the sum of the values of a and b: `sum.Value = a.value+b.value`.

A slightly more complex one is the $\max(a)$ expression. The `max` operator updates its output at each step t of the simulation run by applying the code: `if(a.value > max) max.value = a.value` where `max` is the output value of the operator, and `a.value` is the value of a at time t.

Some operators require more intricate update mechanisms with auxiliary internal variables. Consider the expression $\mathrm{avg}(a)$; the `avg` operator maintains two variables: the number of steps so far, and the sum of all the values of a collected at these step so far. The output value of the expression is the ratio between these two internal variables: `avg.value = sum_of_a / number_of_time_Steps`.

All operators are able to perform such updates of their outputs.

Minimal Updates. To further increase the efficiency of the query evaluation, we distinguish two type of operators: non accumulating ones and accumulating ones. *Non accumulating operators*, such as the $+$ operator do not actually need to perform updates at every step, they just need to be updated at the end of the simulation trace in order to deliver a correct final value. *Accumulating operators* include all the temporal operators, and need to perform some sort of update at each step of the simulation run in order to deliver a correct value at the end of the run.

Some non-accumulating operators might also need to be updated at each step, for instance if their output value is needed by an accumulating operator. We therefore introduce the notion of accumulating expressions and non accumulating expressions. An *accumulating expression* is rooted at an accumulating operator, or is the direct sub-expression of an accumulating expression. *Non-accumulating expressions* are all the remaining sub-expressions.

At each step of the simulation, all operators rooted at accumulating expressions are updated. The non-accumulating expressions are only updated at the end of the simulation run.

Bottom-up Update. A third mechanism that we have incorporated in our query evaluation engine is bottom-up update of the queries expressions. Queries are not structured into trees. They might actually be directed acyclic graphs because some fragment of queries can be shared by several queries. Bottom-up evaluation enables us to update shared fragment of the queries only once, thus gaining effficiency.

Overall Algorithm. When the simulation model and all the queries are instantiated into the engine, a one-shot analysis is performed on the structure of the queries in order to label each sub-expression as accumulating or non-accumulating. All accumulating expressions are then put on an evaluation list, which also captures the order to follow when updating them at each step to comply with the bottom-up order. At each step of the simulation, accumulating expressions are updated following this evaluation list. At the end of the simulation run, non-accumulating expressions are updated once, following a similar bottom-up process.

7   Benchmarking

In order to assess the approach, we took a benchmarking approach based on a set of representative models that were run on the implementations described in the previous section. Our benchmarking addresses our two main contributions:

1. **Expressiveness:** shows that all the risks identified in the cases can easily be captured by the modelling primitives and measurement probes, either by being based on the set of generic probes identified, or by writing case specific probes.
2. **Performance:** shows that running probes does not degrade the performance of the simulation engine a significant amount.

7.1   Benchmark Models

We selected four representative models out of a set of about 20 examples inspired by classic academic cases (these have specific, complex aspects) and other anonymous cases collected in industry. The cases also vary in the level of use of random variables. We describe relevant modelling aspects of each supply chain, together with specific risk issues associated with each model.

First Case: A Simple Assembler Factory. This case, illustrated in Fig. 7, is a simple factory that builds industrial produce using two kinds of parts, Parts A and B. Each part has its supplier which feeds the stocks when they become lower than a given threshold. For part A, the supplier policy is to refill the stock to its maximum capacity. For part B, the policy is the delivery of a fixed amount of material. Part B must be preprocessed before assembly. The factory combines two units of part A with one unit of preprocessed part B, resulting in 80 % of products passing the quality tests. So the assembly process can be represented by a failing single batch process. The goal is to assess if the input stocks are kept within safe limits and are able to cope with production demand.

Second Case: A Beer Game Model. Our second case model is a classical problem called "beer game" [24]. It is a long linear supply chain going from the beer factory to the final retailer, passing by distributors and wholesalers. The beer factory is considered here as a supplier, and single batch processes are used to represent external sources of delay in transport. Intermediary stocks also add extra delays. The continuity of retailing, distribution and wholesaler processes is modelled by conveyor belts. The resulting model is shown in Fig. 1. The goal is to assess where potential bottleneck can occur.

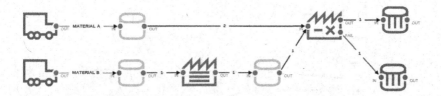

Fig. 7. Model 1: a simple assembler factory.

Third Case: Multiple Suppliers. This case is inspired by a real industrial case, where manufacturing involved three different materials having their own supplier and refill policy, with random delays belonging to a Gaussian probability distribution. 90 % of produce built by a batch in the factory fulfilled the quality requirements. We want to evaluate the effects of different supplying policies in order to ensure the supply chains operate at optimal capacity, whilst minimising the frequency of orders.

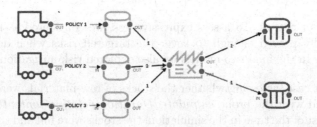

Fig. 8. Model 3: multiple suppliers.

Fourth Case: A Complex Assembly Process. Our last case is inspired by an industrial case in a factory where complex parts are assembled from 3 different materials following a complex process. Two of the parts are preprocessed on factory units that can fail (10 % of failures for the first one, 40 % for the second one). The process is shown in Fig. 9.

7.2 Expressiveness Analysis

We identified a number of basic probes relating to risks directly related to model elements. Such probes are automatically generated. So, for each stock, we generated three different probes for measuring the average and maximum contents of the stock and for verifying whether the stock is full or overflowing. For each process, we generated a probe for measuring the amount of time in which the process was idle or blocked in the simulation.

In addition, the user can specify extra probes for expressing business specific risks that are typically more complex queries in the model. Table 1 summarises some model characteristics like size (suppliers/processes/storages), risks

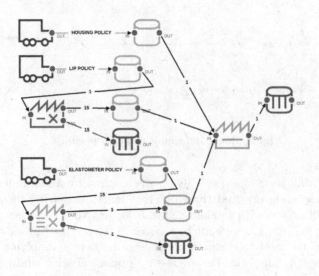

Fig. 9. Model 4: complex assembly process.

and number of probes. To assess expressiveness, we considered a single probe which actually proved enough to cover the targeted risks when dealing with basic probes. In the final two models we also explored risk mitigation strategies.

- In the first case, assessing whether the stocks of raw materials were full could be achieved with the probe $cumulatedDuration(relativeContent(stockA) = 1)$. For most of the time in the simulation, the stocks were full and the assembly process worked at full capacity.
- In the second case, we looked at the relative idle times in the process chain. We noticed that the distribution process, just after the fabrication, is the only one that blocks goods.
- In the third case, we both looked at process idle time (basic probe) and the average contents of stocks using the probe $avg(content(st))$. This helped us discover the best threshold to trigger an order, whilst minimising idle time.
- In the fourth case, a full stock was blocking the production. We mitigated the problem by experimenting with overflowing storage to estimate the right sized storage in order to avoid overflow by using the probe $totalLostByOverflow(lipStorage)$.

7.3 Performance Analysis

A Monte Carlo simulation was run for each case with a time limit of 10000 units and 2000 iterations to be more precise. Table 2 shows the computed average. We performed the benchmarks on an Intel Core i7-4600U CPU at 2.10 GHz with 8 GB of RAM. A single core is currently used. The simulation was triggered from the web interface on the same machine as the server.

Table 1. Benchmarking table for expressiveness.

Name	Size	Risk types	#probes	Comments
M1	9 (2/2/5)	Full stock Process failure	20	Simple manufacturing
M2	17 (1/7/9)	Blocked process Process failure	36	Beer game
M3	8 (3/1/5)	Full stock Process failure Supplier failure	20	Multiple suppliers
M4	14 (3/3/8)	Stock losses	31	Complex part assembly

Table 2. Benchmarking table for performance.

Name	No probes	Std probes	All probes	Overhead
M1	7,3 ms	11,9 ms	12,5 ms	71,2 %
M2	11,3 ms	17,6 ms	17,8 ms	57,5 %
M3	25,8 ms	29,2 ms	34,2 ms	32,6 %
M4	6,8 ms	13,3 ms	13,4 ms	97,1 %

The overhead in the simulation with probes varies from 32,6 % in Model 3 to 97,1 % in Model 4. Model 3 has the longest run time because of the randomness of the supply delays induced by probability distributions associated with suppliers. Model 4 has the shortest run time because the simulation stops at an earlier stage due to a full intermediary stock. In this case the relative overhead is bigger as a result of the DES engine - the load is more efficiently calculated due to the probe's evaluation in the modelling stage.

Globally, overheads are quite acceptable. Some improvements are still possible in the context of integration with a web application, especially in optimising the network requests between the web interface and the simulator, thus making that interface more responsive. The total simulation time allows thousands of simulations to run in a only a few minutes and explore risk mitigation alternatives within an hour.

8 Related Work

A typical risk assessment conducted on a given factory plan is reported in [25], which is based on the Arena simulation tool, featuring DES and Monte Carlo methods as is the case in our work. It stresses the importance of conducting stress-tests using such simulation platforms. Its focus is mainly on the disruption risk, unlike our work which can cope with other classes of risks, like quality for example. Our framework provides an added abstraction layer that can cut down the cost of performing these important stress tests, thus making them achievable by smaller industries.

A similar analysis has been performed on a beer supply chain in [24], whose model was presented in Sect. 7.1. This analysis leads to an evaluation of excessive accumulation in the inventory or back ordering sections. Again, no dedicated tooling was used for representing factories at a higher level, which would lead to higher costs for conducting such an evaluation in an industrial setting. Our tool could cope with using the available primitives.

Another simulation-based risk assessment is reported in [13]. It features an aerospace company with very low production volumes, and leads to the elaboration of a dedicated simulation engine. The engine was first developed with purely deterministic behavior, and then enriched with failure models and stochastic aspects. It was shown to be of great value to the company, despite mainly focusing on disruption risks, it helped the company develop a risk mitigation procedure. Our framework has a similar purpose and tries to propose a compromise between genericity and efficiency.

[26] presents a general framework that combines optimisation and the DES for supporting operational decisions in supply chain networks. Their idea is to iterate between a simulation phase, in which some parameters are estimated, and an optimisation phase that adapts the decision rules for the simulation. Our current work does not cover the minimisation of risk. The tool is rather designed to ease the identification of risk controls by the risk manager. We plan to address optimisation in a later phase, based on the optimisation engines also present in the OscaR framework [4].

9   Conclusions

We presented a Discrete Event Simulation Approach that is supported by a software tool for modelling the supply chain of manufacturing processes with the goal of assessing several kinds of risks on those processes, with a specific focus being placed on procurement risks. This assessment is performed through a simulation engine which uses Monte-Carlo techniques that can also be used to further explore risk mitigation strategies.

The strength of our approach is to support a declarative and easy to use graphical model for representing factory processes and stocks, together with a declarative query language for defining metrics to be measured whilst simulating the behaviour of the modelled system. We could successfully benchmark our approach both from the expressiveness and performance perspectives shown in several typical examples of factories, together with their supply policies.

Further work is still required in order to fully align our approach with the needs of industry. One of our current steps is to validate the tool by putting the tool in the hands of risk managers in a pilot case. We have already identified a number of requests regarding:

– extending the modelling language, e.g. to support the notion of shared resources among processes and have a statistic model of process failures and breakdowns. More specialised processes allowing controlled fork/joins are also required.

- identification of model parameters making easier the manual (and later optimised) exploration of risk mitigation strategies.
- availability of a companion library of specific risks and related probes.
- producing specific reports (e.g. business continuity plans). We have already explored some work in this direction [27].
- possibly creating model refinements and better granularity of the simulation. However our aim is not to capture the full reality but what will help in assessing identified risks.
- finally, parallelisation of the Monte-Carlo simulation engine to obtain better execution times, is needed.

Our framework combining usability, expressiveness and efficiency is an important milestone in our work regarding raising the company awareness, especially in smaller companies, of the need to evaluate their procurement risks and elaborate their supply policies in the most optimal manner. We believe it can be used to manage more general risks. Our design ideas can also be used to improve other risk management tools. Our framework is available online [28] and we plan to make it available open source.

Acknowledgement. This research was conducted under the SimQRi research project (ERA-NET CORNET, Grant No. 1318172). The CORNET promotion plan of the Research Community for Management Cybernetics e.V. (IfU) has been funded by the German Federation of Industrial Research Associations (AiF), based on an enactment of the German Bundestag.

References

1. Deleris, L., Erhun, F.: Risk management in supply networks using Monte-Carlo simulation. In: 2005 Winter Simulation Conference, Orlando, USA (2005)
2. Printz, S., von Cube, J.P., Ponsard, C.: Management of procurement risks on manufacturing processes - survey results (2015). http://simqri.com/uploads/media/Survey_Results.pdf
3. von Cube, J.P., Abbas, B., Schmitt, R., Jeschke, S.: A monetary approach of risk management in procurement. In: 7th International Conference on Production Research Americas' 2014, Lima, Peru, pp. 35–40 (2014)
4. OscaR: OscaR: Scala in OR (2012). https://bitbucket.org/oscarlib/oscar
5. Romeike, F.: Der prozess der risikosteuerung und kontrolle. In: Romeike, F. (ed.) Erfolgsfaktor Risiko-Management, pp. 236–243. Gabler, Wiesbaden (2004)
6. Zsidisin, G.A., Ritchie, B.: Supply Chain Risk: A Handbook of Assessment, Management, and Performance. Springer, New York (2009)
7. Siepermann, M.: Risikokostenrechnung: Erfolgreiche Informationsversorgung und Risikoprävention. Erich Schmidt, Berlin (2008)
8. Sutton, I.: Process Risk and Reliability Management, 2nd edn. Elsevier (2015)
9. Printz, S., von Cube, J.P., Vossen, R., Schmitt, R., Jeschke, S.: Ein kybernetisches modell beschaffungsinduzierter störgößen. In: Exploring Cybernetics - Kybernetik im interdisziplinren Diskurs. Springer Spektrum (2015)
10. Artikis, C., Artikis, P.: Probability Distributions in Risk Management Operations. Springer, London (2015)

11. Zio, E.: The Monte Carlo Simulation Method for System Reliability and Risk Analysis. Springer, London (2013)
12. Gleißner, W.: Quantitative methods for risk management in the real estate development industry. J. Prop. Investment Financ. **30**(6), 612–630 (2012)
13. Finke, G.R., Schmitt, A., Singh, M.: Modeling and simulating supply chain schedule risk. In: 2010 Winter Simulation Conference, Baltimore, USA (2010)
14. Brailsford, S., Churilov, L., Dangerfield, B.: Discrete-Event Simulation and Systems Dynamics for Management Decision Making. Wiley, Chichester (2014)
15. Byong-Kyu, C., Donghun, K.: Modeling and Simulation of Discrete-Event Systems. Wiley (2013)
16. AnyLogic: AnyLogic Multimethod Simulation Software (2015). http://www.anylogic.com
17. Automation, R.: Arena Simulation Software (2015). https://www.arenasimulation.com
18. Siemens: Plant Simulator (2015). http://goo.gl/gH63jw
19. Wampler, D., Payne, A.: Programming Scala. 2nd edn. O'Reilly media (2015)
20. Boostrap: Bootstrap website (2016). http://getbootstrap.com
21. The jQuery Foundation: jQuery website (2016). https://jquery.com
22. ClientIO: JointJS website (2016). http://jointjs.com
23. Scalatra: Scalatra website (2016). http://scalatra.org
24. Klimov, R.A., Merkuyev, Y.A.: Simulation-based risk measurement in supply chains. In: 20th European Conference on Modelling and Simulation (ECMS 2006), Bonn, Germany (2006)
25. Schmitt, A., Singh, M.: Quantifying supply chain disruption risk using Monte Carlo and discrete-event simulation. In: 2009 Winter Simulation Conference, Austin, USA (2009)
26. Almeder, C., Preusser, M., Hartl, R.F.: Simulation and optimization of supply chains: alternative or complementary approaches? In Günther, H.O., Meyr, H. (eds.) Supply Chain Planning, pp. 1–25. Springer, Heidelberg (2009)
27. Arenas, A.E., Massonet, P., Ponsard, C., Aziz, B.: Goal-oriented requirement engineering support for business continuity planning. In: Jeusfeld, M.A., Karlapalem, K. (eds.) ER 2015. LNCS, vol. 9382, pp. 259–269. Springer, Cham (2015). doi:10.1007/978-3-319-25747-1_26
28. SimQRi: Online SimQRi tool (2015). https://simqri.cetic.be

Ramsey's Discrete-Time Growth Model: A Markov Decision Approach with Stochastic Labor

Gabriel Zacarías-Espinoza[1], Hugo Cruz-Suárez[2(✉)],
and Enrique Lemus-Rodríguez[3]

[1] Departamento de Matemáticas, UAM-Iztapalapa, Ave. San Rafael Atlixco 186,
Col. Vicentina, 09340 Mexico City, Mexico
gabrielzaces@hotmail.com

[2] Facultad de Ciencias Físico Matemáticas, Benemérita Universidad Autónoma
de Puebla, Ave. San Claudio y Río Verde, Col. San Manuel, Ciudad Universitaria,
72570 Puebla, Puebla, Mexico
hcs@fcfm.buap.mx

[3] Escuela de Actuaría, Universidad Anáhuac México-Norte, Ave. Universidad
Anáhuac 46, Col. Lomas Anáhuac, 52786 Mexico City, Edo. de México, Mexico
elemus@anahuac.mx

Abstract. In this paper, we study a Markov Decision Process version of the classical Ramsey's Growth model where the evolution of the labor component is assumed to be stochastic. As this is a discrete-time model, it is much easier to study the corresponding long-run behavior of the optimal strategies, as it would be for a continuous version. A set of natural conditions in the Euler Equation context are presented that guarantee a stable long-term behavior of the optimal process.

Keywords: Ramsey's Growth Model · Markov Decision Processes · Dynamic programming · Euler Equation · Stability

1 Introduction

The original model presented by Ramsey in [13] (formulated in communication with the famous economist Keynes) analyzes optimal global saving in a deterministic continuous time setting for a given country. It is no surprise that it was originally solved using Calculus of Variations. Since then, several variants have appeared in the Advanced Macroeconomics literature, but, as Prof. Ekeland points out (one of the leading experts in Mathematical Economics): "To the best of my knowledge and understanding, none of the solutions proposed for solving the Ramsey problem is correct with one exception, of course, Ramsey himself, whose own statement was different than the one which is now in current use [5]". We may safely assume that the intrinsic mathematical difficulty of the Calculus of Variations machinery in the infinite-horizon context introduces an

© Springer International Publishing AG 2017
B. Vitoriano and G.H. Parlier (Eds.): ICORES 2016, CCIS 695, pp. 223–238, 2017.
DOI: 10.1007/978-3-319-53982-9_13

additional layer of complexity to the modeling problem that renders the variants of the model dangerously opaque.

A discrete time version of this problem, on the contrary, allows the straightforward use of Dynamic Programming techniques, and therefore, both researchers and practitioners are able not only to focus on the analysis of the model itself and its properties, but to introduce their own variants addressing related issues of their interest. The deterministic is clearly stated and analyzed in [1, 9, 14].

Ramsey's seminal work on economic growth has been extended in many ways, but, to the best of our knowledge, the study of its random discrete time variants is still in its initial phase. As it has been the case with many stochastic growth models, this allows for a fruitful interaction between economists and mathematician that will lead to better simulations and consequently, to a better understanding of the effects of the random deviations in the growth of an economy and its impact on the population.

In this paper one such random model variant is proposed, where the population (the labor) grows in a stochastic manner. The discrete-time stochastic Ramsey growth problem is stated as a discounted Markov Decision Process (MDP) (see [6, 7]) with total discounted reward as the performance index. The optimal control problem is to determine a savings policy that optimizes the performance criterion. The solution of the optimization problem is analyzed using the Euler Equation (henceforth denoted EE) approach, [3, 4]. Later, the EE is applied to study the ergodic behavior of the stochastic Ramsey growth process.

The fact that the modeling and computational complexity of the present model appear to be reasonable, opens the avenue to the study of further variants of the model, where the population is treated with more detail and the specific impact of varied savings strategies on specific population segments is explicitly dealt with in the mathematical treatment of the model. Hopefully this will lead in the future to a mathematical research school that may relate macro and micro models, where the effect of a policy on specific individual agents (for instance, households) may be quantified. This variants should include more sophisticated and meaningful performance indexes: for instance, utility on consumption, on a closer analysis, depends of flawed philosophical and psychological assumptions.

2   Ramsey's Growth Model

Consider an economy in which at each discrete time t, $t = 0, 1, ...$, there are L_t consumers (population or labor), with consumption c_t per individual, whose growth is governed by the following difference equation:

$$L_{t+1} = L_t \eta_t, \tag{1}$$

it is assumed that initially the number of consumers, L_0, is known. In this case, $\{\eta_t\}$ is a sequence of independent and identical distributed (i.i.d) random variables. The random variable η_t, $t \geq 0$, represents an exogenous shock that affects the consumer population, for example: epidemics, wars, natural disasters, new technology, etc. Then, in this context, it will be supposed that for each $t \geq 0$: $\eta_t > 0$, almost surely, i.e. $P(\eta_t > 0) = 1$.

Remark 1. In the literature of economic growth models is usual to assume that the number of consumers grow very slowly in time, see, for instance, [9,14]. Observe that the model presented in this paper is a first step in an effort to weaken that constraint of the model.

The production function for the economy is given by

$$Y_t = F(K_t, L_t),$$

with K_0 known, i.e. the production Y_t is a function of capital, K_t, and labor, L_t, where the production function, F, is a homogeneous function of degree one, i.e. $F(\lambda x, \lambda y) = \lambda F(x, y)$. The output must be split between consumptions $C_t = c_t L_t$ and the gross investment I_t, i.e.

$$C_t + I_t = Y_t. \tag{2}$$

Let $\delta \in (0, 1)$ be the depreciation rate of capital. Then the evolution equation for capital is given by:

$$K_{t+1} = (1 - \delta)K_t + I_t. \tag{3}$$

Substituting (3) in (2), it is obtained that,

$$C_t - (1 - \delta)K_t + K_{t+1} = Y_t. \tag{4}$$

In the usual way, all variables can be normalized into per capital terms, namely, $y_t := Y_t/L_t$ and $x_t := K_t/L_t$. Then (4) can be expressed in the following way:

$$c_t - (1 - \delta)x_t + K_{t+1}/L_t = F(x_t, 1).$$

Now, using (1) in the previous relation, it yields that

$$x_{t+1} = \xi_t(F(x_t, 1) + (1 - \delta)x_t - c_t),$$

$t = 0, 1, 2, ...,$ where $\xi_t := (\eta_t)^{-1}$.

Define $h(x) := F(x, 1) + (1 - \delta)x$, $x \in X := [0, \infty)$, h henceforth to be identified as the production function. Then, the transition law of the system is given by

$$x_{t+1} = \xi_t(h(x_t) - c_t), \tag{5}$$

$x_0 = x$ known, where $c_t \in [0, h(x_t)]$ and $\{\xi_t\}$ is a sequence of i.i.d. random variables with a density function Δ.

A plan or consumption sequence is a sequence $\pi = \{\pi_n\}_{n=0}^{\infty}$ of stochastic kernel π_n on the control set given the history

$$h_n = (x_1, c_1, \cdots, x_{n-1}, c_{n-1}, x_n),$$

for each $n = 0, 1, \cdots$. Namely, for each $n \geq 1$, π_n is an stochastic kernel if satisfy the following properties:

i. $\pi_n(\cdot \mid h_n)$ is a probability measure on X, for each h_n.
ii. $\pi_n(B \mid \cdot)$ is a random variable, for each $B \in \mathcal{B}(X)$, $\mathcal{B}(X)$ denotes the Borel σ-algebra of X.

The set of all plans will be denoted by Π.

Given an initial capital $x_0 = x \in X$ and a plan $\pi \in \Pi$, the performance index used to evaluate the quality of the plan π is given by

$$v(\pi, x) = \mathbb{E}_x^\pi \left[\sum_{n=0}^{\infty} \alpha^n U(c_n) \right], \tag{6}$$

where $U : [0, \infty) \to \mathbb{R}$ is a measurable function known as utility function and $\alpha \in (0, 1)$ is a discount factor.

The goal of the controller is to maximize utility of consumption on all plans $\pi \in \Pi$, that is:

$$V(x) := \sup_{\pi \in \Pi} v(\pi, x),$$

$x \in X$.

Throughout of this paper the model will be called a Stochastic version of the Ramsey Growth model, in short SRG.

The following assumptions it will be considered in the rest of the document.

Assumption 1. The production function h, satisfies:

(a) $h \in C^2((0, \infty))$.
(b) h is a concave function on X.
(c) $h' > 0$ and $h(0) = 0$.
(d) Let $h'(0) := \lim_{x \downarrow 0} h'(x)$. Suppose that $h'(0) > 1$ and

$$\alpha h'(0) > E[\xi^{-1}]. \tag{7}$$

Remark 2. Observe that, if $x_t = 0$, for some $t \in \{0, 1, 2, ...\}$ then $x_k = 0$ for each $k \geq t$. This fact is a consequence of (5) and Assumption 1 (c). Consequently the state zero is an absorption state.

Assumption 2. The utility function U satisfies:

(a) $U \in C^2((0, \infty), \mathbb{R})$, with $U' > 0$ and $U'' < 0$.
(b) $U'(0) = \infty$ and $U'(\infty) = 0$
(c) There exists a function ϑ on S such that $E[\vartheta(\xi)] < \infty$, and

$$|U'(h(s(h(x) - c)))h'(s(h(x) - c))s\Delta(s)| \leq \vartheta(s), \tag{8}$$

$s \in S$, $c \in (0, h(x))$.

Remark 3. Observe that in Assumption 2 is not considered the Inada condition in zero. In the literature, it is known to have the rather unrealistic implication that each unit of capital must be capable of producing an arbitrarily large amount of output with a sufficient amount of labor [8].

3 Dynamic Programming Approach

In this section it will be presented an analysis of the optimization problem introduced in Sect. 2. Dynamic Programming approach have been used to study different type of problems and in various context. In particular have been applied to Markov Decision Processes (MDP).

SRG can be identified as a MDP. In this case, the space of states is $X := [0, \infty)$, the admissible action space is $A(x) := [0, h(x)], x \in X$, in consequence, the action space is $A := \bigcup_{x \in X} A(x) = [0, \infty)$. The transition law is given by the stochastic kernel, defined as

$$Q(B \mid x_t = x, a_t = c) = \Pr(x_{t+1} \in B \mid x_t = x, a_t = c) \tag{9}$$

$$= \int_B w(x, y, c) dy, \tag{10}$$

with $B \in \mathcal{B}(X)$, $\mathcal{B}(X)$ denotes the Borel sigma algebra of X, where the function $w : [0, \infty)^3 \to [0, \infty)$ is defined as:

$$w(x, y, c) := \Delta\left(\frac{y}{h(x) - c}\right) \frac{1}{h(x) - c}, \tag{11}$$

for $x, y \in X$, $c \in [0, h(x))$ and Δ is the density function of the sequence $\{\xi_t\}$. Define $\mathbb{K} := \{(x, c) \mid x \in X, c \in A(x)\}$. Finally, the reward-per-stage function is identified as the utility function, $U : X \to [0, \infty)$, defined in the previous section. Then the model is referred as the quintuplet: $\mathcal{M} := (X, A, \{A(x) : x \in X\}, Q, U)$.

As it was mentioned above, a plan is a sequence $\pi = \{\pi_n\}_{n=0}^{\infty}$ of stochastic kernel defined on A given the history of the process. Furthermore, it is assumed that $\pi_n(A(x_n) \mid h_n) = 1$, $n = 0, 1, \cdots$, this assumption guarantee that in each decision epoch, it is possible to choose an admissible action. A particular class in Π is the class of *stationary plans*,

$$\mathbb{F} := \{f : X \to A \mid f(x) \in [0, h(x)], \text{ for all } x \in X\}.$$

In this case, a stationary plan $\pi = (f, f, ...)$ is denoted by f.

Under Assumptions 1 and 2, for each $x \in X$, it follows that:

(a) The *optimal value function* V satisfies the *following equation (optimality equation)*

$$V(x) = \sup_{c \in A(x)} \left\{ U(c) + \alpha \int_0^{\infty} V(y) w(x, y, c) dy \right\}. \tag{12}$$

(b) There exists and *optimal stationary policy* $f \in \mathbb{F}$ such that

$$V(x) = U(f(x)) + \alpha \int_0^{\infty} V(y) w(x, y, f(x)) dy.$$

(c) $v_n(x) \to V(x)$ when $n \to \infty$, where v_n is defined by

$$v_n(x) = \sup_{c \in A(x)} \left\{ U(c) + \alpha \int v_{n-1}(y)w(x,y,c)dy \right\},$$

with $v_0(x) = 0$.

A proof of the statement (a), (b) and (c) can be consulted in [6].

Remark 4. The functions, v_n, $n \geq 0$, defined on (c) are known as *value iteration functions*, [6].

4   Stochastic Euler Equation

In this section it will be presented a functional equation, which characterize the optimal value function. In the literature of MDP, this functional equation is known as Euler Equation (EE), [4]. The validity of EE is guaranteed due to properties of differentiability of the optimal value function and the optimal policy, [2]. Then, it just is necessary to verified that the optimal policy is interior, according to Theorem 3.3 in [3], this fact is verified in the following result.

Lemma 1. *The optimal plan f satisfies that $f(x) \in (0, h(x))$, for each $x > 0$.*

Proof. Let $x > 0$ fixed, if the optimal policy is $f(\cdot) \equiv 0$, then

$$V(x) = v(0,x) = \frac{U(0)}{1-\alpha},$$

where v is defined in (6).

Since U and h are strictly increasing (see Assumptions 1 and 2), it is obtained that

$$V(x) = \frac{U(0)}{1-\alpha} < U(h(x)) + \frac{\alpha}{1-\alpha} U(0),$$

but this is a contradiction, given that

$$v(h,x) = U(h(x)) + \frac{\alpha}{1-\alpha} U(0).$$

On the other hand, if $h \in \mathbb{F}$ is the optimal policy, then

$$V(x) = v(h,x)$$
$$= U(h(x)) + \frac{\alpha}{1-\alpha} U(0).$$

Let $g : [0, h(x)] \to \mathbb{R}$ be a function defined as

$$g(c) := U(c) + \alpha E[U(h(\xi(h(x) - c)))] + \frac{\alpha^2}{1-\alpha} U(0).$$

Observe that g is continuous and a strictly concave function. Then, there exists an unique $\bar{c} \in [0, h(x)]$, which maximizes to g. If $\bar{c} \neq h(x)$, then

$$V(x) \geq g(\bar{c}) > g(h(x)) = V(x),$$

which is impossible. Therefore $\bar{c} = h(x)$.

Now, Assumptions 1 and 2 imply that if $c \in (0, h(x))$,

$$g'(c) = U'(c) - \alpha E[U'(h(\xi(h(x) - c)))h'(\xi(h(x) - c))\xi],$$

it follows that

$$\lim_{c \to h(x)} g'(c) = -\infty.$$

Therefore, there exists $\tilde{c} \in (0, h(x))$ such that $g'(\tilde{c}) < 0$. This implies that g is decreasing in $[\tilde{c}, h(x)]$ which $h(x)$ can not be the maximizer, i.e. it is a contradiction.   □

A consequence of Lemma 1 is the following result. The proof of Theorem 1 it follows of Lemma 5.2 and Lemma 5.6 in [3].

Theorem 1. *Under Assumptions 1 and 2, it follows that:*

(a) $V \in C^2((0, \infty), \mathbb{R})$ and the optimal plan $f \in C^1((0, \infty))$.
(b) The value iteration functions satisfies

$$\frac{v_n'(x)}{h'(x)} = \alpha E\left\{ v_{n-1}'\left[\xi\left(h(x) - U'^{-1}\left(\frac{v_n'(x)}{h'(x)} \right) \right) \right] \xi \right\},$$

for each $x > 0$, where $U'^{-1}(\cdot)$ is the inverse of $U'(\cdot)$.
(c) The optimal plan f satisfies the following Euler equation:

$$U'(f(x)) = \alpha E[U'(c^*(x))h'(\xi(h(x) - f(x)))\xi],$$

for each $x > 0$, where $c^*(x) := f(\xi(h(x) - f(x)))$.

Remark 5. Observe that if $f \in \mathbb{F}$ satisfies (1) and

$$\lim_{n \to \infty} \alpha^n E_x^f [h'(x_n)U'(f(x_n))x_n] = 0,$$

then f is an optimal plan.

4.1 Auxiliary Tools About Stability

It is known that if $f \in \mathbf{F}$ is the optimal plan then $\{x_n\}$ is a Markov chain with transition kernels given in (10). Thus, in this section some definitions and results of ergodic theory for Markov chains are presented, which are taken of reference [10].

Let us consider a Markov chain $\{x_n\}$ defined in a space X, with transition Kernels $Q(\cdot|\cdot)$ and a weight function, w, $w : X \to [1, \infty)$. Let $\mathbb{B}_w(X)$ be the space of measurable and w-bounded function on X with norm $\|\cdot\|_w$ defined as

$$\|u\|_w := \sup_{x \in X} \frac{|u(x)|}{w(x)}, \tag{13}$$

for a measurable function u on X. Let φ be a signed measure defined on the Borel σ-algebra of X, denoted by $\mathcal{B}(X)$. Then, for each $u \in \mathbb{B}_w(X)$, φu denotes the function defined as

$$\varphi u := \int u(y)\varphi(dy).$$

In terms of the stochastic kernel Q, Qu has the form

$$Qu(x) = \int u(y)Q(dy|x),$$

and the n-th transition kernel is

$$Q^n(B|x) = \int Q(B|y)Q^{n-1}(dy|x), B \in \mathcal{B}(X),$$

for $n \geq 1$ and $Q^0 := \delta_x$, where δ_x is Dirac's measure on $x \in X$.

Definition 1. *Let P be a probability measure on $\mathcal{B}(X)$. The measure P is invariant (m.p.i) with respect to the chain $\{x_n\}$, if $QP = P$, where*

$$QP(B) := \int Q(B|y)P(dy), B \in \mathcal{B}(X).$$

Define for each measure φ on $\mathcal{B}(X)$, the w-norm as,

$$\|\varphi\|_w := \sup_{\|u\|_w \leq 1} |\varphi u|.$$

Definition 2. *The Markov chain $\{x_n\}$ is w-geometrically ergodic if there exists a probability measure P and non-negative constants R and ρ, $\rho < 1$, such that for each $n = 0, 1, \ldots$*

$$\|Q^n - P\|_w \leq R\rho^n.$$

Definition 3. *The Markov chain $\{x_n\}$ is φ-irreducible, if there exists a measure φ on $\mathcal{B}(X)$ such that, if $\varphi(A) > 0$ then $\Pr[X_n \in A$ for some $n \in \mathbb{N}] > 0$.*

Definition 4. *Given a set $C \in \mathcal{B}(X)$. C is a small set with respect to the Markov chain $\{x_n\}$, if there exist a finite measure μ on $\mathcal{B}(X)$ and $n \in \mathbb{N}$, such that for each $x \in C$ and $B \in \mathcal{B}(X)$:*

$$Q^n(B|x) \geq \mu(B).$$

Definition 5. *The Markov chain $\{x_n\}$ is strongly aperiodic if there exists a small set C such that $\mu(C) > 0$.*

The following result is stated and proved in [10], Theorem 16.1.2, p. 395.

Theorem 2. *Suppose that the Markov chain $\{x_n\}$ is φ-irreducible, strongly aperiodic and there are a small set C and constants $\beta < 1$ and b such that for each $x \in X$*

$$Qw(x) \leq \beta w(x) + bI_C(x).$$

Then there exists a unique m.p.i. P such that the chain is w-geometrically ergodic.

Definition 6. *A Lyapunov function on a topological space S is a real-valued non-negative function W defined on Y such that each level set is precompact, that is, for each $a \in \mathbb{R}$ the set $\{x : W(x) \leq a\}$ has a compact closure.*

Theorem 2 implies that the sequence of random variables of the Markov chain converges in distribution to the random variable generated by the m.p.i. (see Definition 1). From that, we get convergence in L^1, as a consequence of the following lemma, the proof of Lemma 2 can be consulted in [12] (Theorem 5.9, p. 224).

Lemma 2. *Let $\{X_n\}$ be a sequence of random variables that converge in distribution to X. If for some $p > 0$, the sequence $\{|X_n|^p\}$ is uniformly integrable then*

$$E[|X_n|^p] \rightarrow E[|X|^p],$$

when $n \rightarrow \infty$. If $p \geq 1$, then $E[X_n] \rightarrow E[X]$ when $n \rightarrow \infty$.

5   Stability for the SRG Model

In the spirit of Nishimura and Stachurski (see [11]), we will use density functions on $(0, \infty)$ and with the aid of the Euler Equation as presented in Theorem 1, convergence on this model will be established.

If $f \in \mathbb{F}$ is the stationary optimal plan for the RGU model, the *stochastic optimal process* is given by

$$x_{t+1} = \xi_t(h(x_t) - f(x_t)),$$

$t = 0, 1, 2, \ldots$, $x_0 = x \in X = [0, \infty)$, known.

It will be consider the non-trivial case, i.e. it will be assumed that $x_0 = x > 0$. In this case the stochastic kernel is given by

$$Q(B \mid x, f(x)) = \int_{\{s \mid s(h(x)-f(x)) \in B\}} \Delta(s)ds, \; x > 0,$$

for each $B \in \mathcal{B}(X)$, it will be assumed that the density Δ is continuous and positive, and consequently the kernel $Q(\cdot \mid x, f(x))$ density is determined by

$$q_1(y \mid x, f(x)) := w(x, y, f(x)),$$

for each $x, y \in (0, \infty)$. Observe that $h - f$ is strictly positive on $(0, \infty)$, due to $f(x) \in (0, h(x))$ (see Lemma 1).

Then, given $x > 0$, by induction we get that for each $t \geq 1$ and $x_t \in X$ the corresponding density q_t, which is determined by

$$q_t(y) = \int q_1(y \mid x, f(x)) q_{t-1}(x) dx.$$

Define for $A \in B(X)$ the measure

$$\Xi(A) := \int_A \Delta(s) ds.$$

Lemma 3. *The optimal process $\{x_n\}$ of RGU is Ξ-irreducible and strongly aperiodic.*

Proof. Let $B \in \mathcal{B}(X)$ such that $\Xi(B) > 0$ and $x > 0$. Then, we know that $h(x) - f(x) > 0$ because f is an interior point in the corresponding interval. Moreover,

$$\Pr(x_1 \in B) = \int_X I_B(h(x) - f(x)s)\Delta(s)ds,$$

where I_B denotes the indicator function of the set B. As Δ is positive, it follows that $\Pr(x_1 \in B) > 0$. Consequently, the optimal process $\{x_n\}$ is irreducible (see Definition 3).

Now, it will proved that the stochastic optimal process is strongly aperiodic. Consider $a, b \in (0, \infty)$, with $a < b$ and define $C := [a, b]$. Observe that $h - f$ is an increasing function. To prove this fact, suppose that $h - f$ is non-increasing on $(0, \infty)$. Then, for $x, y \in (0, \infty)$, with $x < y$, the following inequality holds

$$h(y) - f(y) \leq h(x) - f(x).$$

This imply that

$$U'(h(x) - f(x))h'(x) \leq U'(h(y) - f(y))h'(y), \tag{14}$$

due to Assumptions 1 and 2. Applying the Envelope Formula (see [2]) in (14), it follows that $V'(x) \leq V'(y)$, which is a contradiction, due to the derivative of the optimal value function, V, is a decreasing function, since $V \in C^1$ is a concave function. Consequently $h - f$ is an increasing function, this fact, imply that for each $x \in C$:

$$0 < h(a) - f(b) \leq h(x) - f(x) \leq h(b) - f(b).$$

Furthermore, due to Δ is a continuous and positive function, it follows that

$$m := \inf_{(x,s) \in C \times C} \Delta\left(\frac{s}{h(x) - f(x)}\right) \frac{1}{h(x) - f(x)} > 0.$$

Therefore, for μ a measure on $B \in \mathcal{B}(X)$ defined by

$$\mu(B) := m \int_B I_C(x)dx,$$

it is obtained that

$$Q(B \mid x) \geq \mu(B).$$

Concluding that C is a small set (see Definition 4). Since, $\mu(C) > 0$, it is concluded that the process is strongly aperiodic (see Definition 5).   □

Define W for $x \in (0, \infty)$ as

$$W(x) := [U'(f(x))h'(x)]^{1/2} + x^p + 1, \tag{15}$$

with $p > 1$.

Lemma 4. *Function W is a Lyapunov function.*

Proof. Consider $a \in \mathbb{R}$ and

$$N_a := \{x \in (0, +\infty) \mid W(x) \leq a\}$$

Suppose that $a \leq 1$, as $W(x) > 0$, for each $x > 0$, then $N_a = \emptyset$ and hence its closure is compact.

On the other hand, if $a > 1$ and $\{x_n\}$ is a sequence in N_a such that $x_n \to x$, due to the continuity of W it follows that $x \in N_a$ and consequently, N_a is a closed set. Due to Assumptions 1 and 2 and the definition of W it is immediate that N_a is bounded and hence, compact, and trivially, with compact closure. Since a is arbitrary, it is concluded that W is a Lyapunov function (see Definition 6)   □

Lemma 5. *Let $w_1(x) := [U'(f(x))h'(x)]^{1/2}$, for $x > 0$ and suppose that (7) is satisfied, then there exist constants λ_1 and b_1 such that $\lambda_1 \in (0, 1)$ and for each $x \in (0, \infty)$*

$$\int_0^\infty w_1((h(x) - f(x))s)\mu(ds) \leq \lambda_1 w_1(x) + b_1. \tag{16}$$

Proof. Consider the function w_1 defined on $(0, \infty)$, then by Cauchy-Schwartz inequality, it is obtained that

$$\int_0^\infty w_1(u(x)s)\mu(ds) \leq \left(\int_0^\infty U'(u(x)s)h'(u(x)s)s\mu(ds) \right)^{1/2} \left(\int_0^\infty 1/s\mu(ds) \right)^{1/2},$$

where $u(x) = h(x) - f(x), x > 0$. Applying the EE (see Theorem 1 (c)) to the last display, it yields that

$$\int_0^\infty w_1(u(x)s)\mu(ds) \leq (U'(f(x)))^{1/2} \left(E\left[\xi^{-1}\right]/\alpha \right)^{1/2}$$

$$= w_1(x) \left(E\left[\xi^{-1}\right]/h'(x)\alpha \right)^{1/2}$$

Since (7) holds, there exist $\delta > 0$ and $\lambda \in (0,1)$ such that

$$E\left[\xi^{-1}\right]/\alpha h'(x) < \lambda,$$

if $x < \delta$.

In consequence, for $x < \delta$ the following inequality holds

$$\int_0^\infty w_1(u(x)s)\mu(ds) \leq \lambda w_1(x), \tag{17}$$

On the other hand, as U' and h' are decreasing functions and $h - f$ is an increasing function, then the function defined as $U'(u(x))h'(x)$ is a decreasing function on $(0,\infty)$. Then, by (17), it is obtained that

$$\int_0^\infty w_1((h(x) - f(x))s)\Delta(s)ds \leq b_1, \tag{18}$$

if $x \geq \delta$, where

$$b_1 := \left(\int_0^\infty U'(f((u(\delta))s))h'((u(\delta))s)s\mu(ds)\right)^{1/2} \left(E\left[\xi^{-1}\right]\right)^{1/2} < \infty.$$

The result follows from (17) and (18).

□

Lemma 6. *Let $w_2(x) = x^p$, then, under Assumptions 1 and 2 there exist constants λ_2 and b_2 such that $\lambda_2 \in (0,1)$ and for each $x \in (0,\infty)$*

$$\int_0^\infty w_2((h(x) - f(x))s)\Delta(s)ds \leq \lambda_2 w_2(x) + b_2. \tag{19}$$

Proof. Let $\gamma \in (0,1)$ be such that $\gamma^p E[\xi^p] < 1$, and, given Assumption 1-(d) the existence of a constant d such that $h(x) < \gamma x$ is guaranteed.

Consequently, if $x > d$ then

$$\int_0^\infty ((h(x) - f(x))s)^p \Delta(s)ds \leq \gamma^p E[\xi^p]x^p.$$

On the other hand, if $x \in (0,d]$ then $f(x) \in (0,h(d))$ and hence

$$\int_0^\infty w_2((h(x) - f(x))s)\Delta(s)ds = \int_0^\infty ((h(x) - f(x))s)^p \Delta(s)ds$$

$$\leq (h(d))^p E[\xi^p].$$

Taking $\lambda_2 := \gamma^p E[\xi^p]$ and $b_2 := (h(d))^p E[\xi^p]$, Lemma 6 follows. □

Corollary 1. *Let W be the function defined in (15). Then there exist constants λ and b such that $\lambda \in (0, 1)$ and for each $x \in (0, \infty)$*

$$\int_0^\infty W((h(x) - f(x))s)\Delta(s)ds \leq \lambda W(x) + b. \tag{20}$$

Proof. Since w_1 and w_2 satisfy (16) and (19), respectively. Then W satisfies (20) with $\lambda = \max\{\lambda_1, \lambda_2\}$ and $b := b_1 + b_2 + 1$. $\square$

Lemma 7. *Let $\lambda \in (0, 1)$ and $b \in \mathbb{R}$ as Corollary 1. Then, there exists a compact subset C of X such that*

$$\int_0^\infty W((h(x) - f(x))s)\Delta(s)ds \leq \beta W(x) + bI_C(x), \tag{21}$$

with $\lambda < \beta < 1$.

Proof. Let $\beta \in (\lambda, 1)$. Now observe that there exists a compact set $C \subseteq (0, \infty)$ such that $W(x) \geq b/(\beta - \lambda)$ (see Definition 6). Then, by Corollary 1, for each $x \in C$

$$\int_0^\infty W((h(x) - f(x))s)\Delta(s)ds \leq \lambda W(x) + b$$

$$\leq \beta W(x) + bI_C(x),$$

On the other hand for $x \notin C$ (21) holds, as follows

$$\int_0^\infty \frac{W((h(x) - f(x))s)\Delta(s)ds}{W(x)} \leq \lambda + \frac{b}{W(x)} \leq \beta.$$

$\square$

Theorem 3. *For the optimal process in the SRG model there exists a unique m.p.i. P and the stochastic process is W-geometrically ergodic.*

Proof. Due to Lemma 3 the optimal process is Δ-irreducible and strongly aperiodic. Moreover, there exists a compact subset C of X that is small and such that for each $x \in X$ inequality (21) holds. Therefore, from Theorem 2 the result follows. $\square$

Theorem 4. *The SRG optimal process converges in L^1 to a random variable with probability measure P given in Lemma 4.*

Proof. Let $x_0 = x \in X$, $\{x_n\}$ be the optimal process. By Lemma 6 it is known that Lyapunov condition (19) holds for $\{x_n\}$. Hence, for $n = 0, 1, \ldots$,

$$E[x_{n+1}^p \,|\, x_n] \leq \lambda_2 x_n^p + b_2, \tag{22}$$

where λ_2 and b_2 are the constants in Lemma 6. Iterating (and applying standard conditional expectation properties) (22) for $n = 0, 1, \ldots$, it follows that for each $n \in \mathbb{N}$:

$$E[x_{n+1}^p] \leq x_0^p + \frac{b_2}{1 - \lambda_2} < \infty,$$

hence

$$\sup_n E[x_n^p] < \infty.$$

Moreover, since x_n is almost surely positive, by Theorem 4.2 in [12], it follows that the optimal process is uniformly integrable. Furthermore, by Theorem 3 it is known that the sequence $\{x_n\}$ converges in distribution to the m.p.i. P and from Lemma 2 the desired result follows.   □

6   Examples: Cobb-Douglas Utility

Consider the following utility function:

$$U(c) = \frac{b}{\gamma} c^\gamma,$$

for $c > 0$, where $b > 0$ and $\gamma = 1/3$. The transition law is determined by

$$x_{t+1} = \xi_t (x_t - a_t),$$

$a_t \in [0, x_t]$, $t = 0, 1, 2, \ldots$, $x_0 = x \in (0, \infty)$. Observe that in this case the production function $h(x) = x$, $x \in (0, \infty)$. Suppose that $\{\xi_t\}$ is a sequence of i.i.d. random variables independent of x_0. Let ξ a generic element of $\{\xi_t\}$ and consider that ξ with log-normal distribution with mean $3/2$ and variance 1. Then:

$$\mu_\gamma := E[\xi^\gamma] = e^{5/9}$$

it is easy to see that $0 < \alpha\mu_\gamma < 1$, where the discount factor $\alpha < e^{5/9}$. Moreover

$$E\left[\xi^{-1}\right] = e^{-1}$$

and Assumption 1 (d) holds.

Remark 6. Assumption 1 (d) holds for a log-normal distribution if and only if $\sigma^2 < 2\mu$ where μ and σ^2 are mean and variance, respectively.

Define $\delta := (\alpha\mu_\gamma)^{1/(\gamma-1)}$. It is shown in [3] that

$$f(x) := \left(\frac{\delta - 1}{\delta}\right) x, \qquad (23)$$

$x \in X$, is the optimal plan.

Consider the process corresponding to the optimal plan:

$$x_{t+1} = \xi_t \left(x_t - f(x_t) \right)$$

$t = 0, 1, 2, \ldots$, $x_0 = x \in X$; easy calculations show that

$$x_{t+1} = \frac{\xi_t x_t}{\delta}$$

iterating this last equation we get

$$x_{t+1} = \frac{x}{\delta^t} \prod_{i=0}^{t-1} \xi_i.$$

Taking the expectation and using the independence of $\xi_0, \xi_1, \ldots, \xi_{t-1}$, it yields that

$$E\left[x_{t+1}\right] = \frac{x}{\mu} \left(\frac{\mu}{\delta} \right)^t,$$

where $\mu := E\left[\xi\right] = e^2$.

Finally, if $\mu < \delta$ then $E[X_n]$ increasing indefinitely with respect the time; if $\mu > \delta$ then $E[X_n]$ decreasing to zero; if $\mu = \delta$ then $E[X_n] = x/\mu$.

7   Conclusions

It is clear that a natural next step would be to numerically simulate the behavior of this model and find approximations of the stationary distribution. In particular, in the case of concrete utility functions as the Cobb-Douglas, it would be very interesting to compare the dependence of the stationary distribution with respect to its parameters. As it may be easily be confirmed browsing through the literature stochastic growth models present a series of interesting mathematical challenges and hence research opportunities. On the other hand, as many of those models either consider the underlying human populations from a too abstract and macro point of view, or hyper-simplify their well-being through the use of these very same utility functions, it is would be worth stressing the interdisciplinary problems that arise in this case.

How could we introduce more philosophically and psychologically robust performance indexes?

How could we explicitly deal with individual agents, i.e., people, households, small and medium enterprises?

These questions are related, as macro performance indexes as utility on the aggregated consumption of an entire population fail both to describe the real, phenomenological well being (or in some cases "bad being") of the agents or its enormous intrinsic heterogeneity: as these models are usually simplified from a demographic point of view.

As a matter of fact, we may even need to dispute the very notion of growth at all costs, that is implicitly embedded in the classical models, as many thinkers start to worry about the toxic effect on human beings brought this growth mentality that for instance allow of ignore the corporate-driven growth of very large sectors of the economy (optimizing shareholder return appears to have negative effects on the workers and consumers).

In conclusion, if there is much to be done from the purely mathematical point of view regarding growth models, there is much more to be done from both the interdisciplinary modeling and mathematical perspectives if we want to build new models that explicitly deal with the well-being of the individual agents of the corresponding population.

Interestingly enough, it is to be expected that such new models may be initially built as multicriteria Markov Decision Processes.

References

1. Brida, J.G., Cayssials, G., Pereyra, J.S.: The discrete Ramsey model with decreasing population growth rate. Dyn. Continuous Discrete Impulse Syst. Ser. B Appl. Algorithms **22**, 97–115 (1981)
2. Cruz-Suárez, H., Montes-de-Oca, R.: An envelope theorem and some applications to discounted Markov decision processes. Math. Methods Oper. Res. **67**, 299–321 (2008)
3. Cruz-Suárez, H., Montes-de-Oca, R., Zacarías, G.: A consumption-investment problem modelled as a discounted Markov decision process. Kybernetika (Prague) **47**, 909–929 (2011)
4. Cruz-Suárez, H., Zacarías-Espinoza, G., Vázquez-Guevara, V.: A version of the Euler equation in discounted Markov decision processes. J. Appl. Math. **2012**, 16 (2012)
5. Ekeland, I.: From Frank Ramsey to René Thom: a classical problem in the calculus of variations leading to an implicit differential equation. Discrete Continuous Dyn. Syst. **28**, 1101–1119 (2010)
6. Hernández-Lerma, O., Lasserre, J.B.: Discrete-time Markov Control Processes. Springer, New York (1996)
7. Jaśkiewicz, A., Nowak, A.S.: Discounted dynamic programming with unbounded returns: application to economic models. J. Math. Anal. Appl. **378**, 450–462 (2011)
8. Kamihigashi, T.: Almost sure convergence to zero in stochastic growth models. Econ. Theor. **29**, 231–237 (2006)
9. Le Van, C., Rose-Anne, D.: Dynamic Programming in Economics. Kluwer Academic Publishers, Dordrecht (2003)
10. Meyn, S., Tweedie, R.L.: Markov Chains and Stochastic Stability. Cambridge University Press, Cambridge (2009)
11. Nishimura, K., Stachurski, J.: Stability of stochastic optimal growth models: a new approach. J. Econ. Theor. **122**, 100–118 (2005)
12. Peligrad, M., Gut, A.: Almost-sure results for a class of dependent random variables. J. Theor. Probab. **12**, 87–104 (1999)
13. Ramsey, F.P.: A mathematical theory of saving. Econ. J. **38**, 543–559 (1928)
14. Sladký, K.: Some remarks on stochastic version of the Ramsey growth model. Bull. Czech Econometric Soc. **19**, 139–152 (2012)

An Investigation of Heuristic Decomposition to Tackle Workforce Scheduling and Routing with Time-Dependent Activities Constraints

Wasakorn Laesanklang[✉], Dario Landa-Silva, and J. Arturo Castillo-Salazar

School of Computer Science, ASAP Research Group, The University of Nottingham, Jubilee Campus, Wollaton Road, Nottingham NG8 1BB, UK
{psxwl3,dario.landasilva}@nottingham.ac.uk, jacastillo.salazar@gmail.com

Abstract. This paper presents an investigation into the application of heuristic decomposition and mixed-integer programming to tackle workforce scheduling and routing problems (WSRP) that involve time-dependent activities constraints. These constraints refer to time-wise dependencies between activities. The decomposition method investigated here is called repeated decomposition with conflict repair (RDCR) and it consists of repeatedly applying a phase of problem decomposition and sub-problem solving, followed by a phase dedicated to conflict repair. In order to deal with the time-dependent activities constraints, the problem decomposition puts all activities associated to the same location and their dependent activities in the same sub-problem. This is to guarantee the satisfaction of time-dependent activities constraints as each sub-problem is solved exactly with an exact solver. Once the assignments are made, the time windows of dependent activities are fixed even if those activities are subject to the repair phase. The paper presents an experimental study to assess the performance of the decomposition method when compared to a tailored greedy heuristic. Results show that the proposed RDCR is an effective approach to harness the power of mixed integer programming solvers to tackle the difficult and highly constrained WSRP in practical computational time. Also, an analysis is conducted in order to understand how the performance of the different solution methods (the decomposition, the tailored heuristic and the MIP solver) is affected by the size of the problem instances and other features of the problem. The paper concludes by making some recommendations on the type of method that could be more suitable for different problem sizes.

Keywords: Workforce scheduling and routing problem · Time-dependent activities constraints · Mixed integer programming · Problem decomposition.

1 Introduction

This paper applies Repeated Decomposition with Conflict Repair (RDCR) on a mixed integer programming model to tackle a Workforce Scheduling and Routing

© Springer International Publishing AG 2017
B. Vitoriano and G.H. Parlier (Eds.): ICORES 2016, CCIS 695, pp. 239–260, 2017.
DOI: 10.1007/978-3-319-53982-9_14

Problem (WSRP) with time-dependent activities constraints. The WSRP refers to assigning employees with diverse skills to a series of visits at different locations. A visit requires certain skills so that the tasks or activities can be performed. The problem has become important especially in the recent years because the number of businesses using a mobile workforce is growing [1]. These businesses usually provide services to people at their home. Examples are home care [2,3], home healthcare [4], security patrol services [5,6], technician scheduling [7–9], etc. In this type of scenarios, a mobile workforce must travel from its base to visit multiple locations to deliver services.

Time-dependent activities constraints refer to the case in which visits are time-wise related, such feature in WSRP was discussed in [10]. There are five types of time-dependent activities constraints: synchronisation, overlap, minimum difference, maximum difference and minimum-maximum difference. There have been attempts to use mathematical programming to find optimal solutions for WSRP [10,11]. The problems are usually formulated as mixed integer programs (MIP) and implemented as a network flow problem. However, solving the problem using mathematical programming solvers requires very high computational time. Such solvers are able to find optimal solutions for only small instances and only sometimes feasible solution can be found within 4 hours. It has also been shown that when solving larger instances (e.g. more than 150 visits), it is often not possible to find optimal solutions due to the computer memory being exhausted. A constructive greedy heuristic (GHI) was proposed to solve WSRP with time-dependent activities constraints [12]. That algorithm provided better solutions than the mathematical programming solver when the number of visits is more than 100. In addition, other solution methods such variable neighbourhood search [13] and greedy constructive heuristics [14] have also been applied to WSRP instances with time-dependent activities constraints.

In the literature, there are works applying mathematical programming solvers within a decomposition approach to solve real-world problems. A decomposition method breaks a problem into smaller parts which are easier to solve. A problem can be decomposed by exact or heuristic approaches. An example of exact decomposition is Dantzig-Wolfe decomposition where all possible assignment combinations can be generated [15]. This approach may require high computational times to achieve a good solution. Heuristic decomposition generates subproblems by splitting the full problem using some heuristic procedure to solve each sub-problem and integrate the partial solutions into a solution to the full problem. This usually means that heuristic decomposition does not guarantee optimality in the overall solution. An example of heuristic decomposition method is the Geographical Decomposition with Conflict Avoidance (GDCA) proposed in [16,17] to tackle a home healthcare scheduling problem. Those home healthcare instances tackled with GDCA had a fixed time for the visits instead of a time window and had no time-dependent activities constraints. The GDCA technique decomposed a problem by geographical regions resulting in several sub-problems which then are tackled individually. The GDCA method was capable of finding a feasible solution even for instances with more than 1,700 clients.

Other related heuristic decomposition methods using some form of clustering have been presented in the literature. For example, a large vehicle routing problem was decomposed into various clusters of customers assigned to a vehicle in [18].

This paper applies a Repeated Decomposition with Conflict Repair (RDCR) approach to a varied set of WSRP instances that involve time-dependent activities constraints. In general, RDCR decomposes a problem into sub-problems which then are individually solved with a mathematical programming solver. A sub-problem solution gives a path or sequence of visits for each employee. However, since an employee may be used in several sub-problems, this can lead to having *conflicting paths*, i.e. different paths that are assigned to the same employee. Another type of conflict are *conflicting assignments*, i.e. visits overlapping in time assigned to the same employee. Avoiding conflicting assignments within a path is guaranteed by the mathematical programming model. However, conflicting paths can arise because sub-problems are individually solved and the available workforce is shared among sub-problems. Therefore, conflicting paths need to be resolved by a conflict repair process described later in this paper. The stage of problem decomposition and sub-problem solving is followed by the stage of conflict repair. These two stages are repeatedly applied as part of the RDCR method until no more visits can be assigned in the current solution. This paper compares the solution quality from the proposed decomposition method to the results produced by the greedy constructive heuristic (GHI) presented in [12]. Moreover, this paper also conducts an in-depth analysis of the performance by the RDCR and GHI methods in respect of the problem features. This analysis aims to identify the types of problem instances in which each of the methods performs better. This will contribute to a better understanding of what type of approach is expected to be more successful according to the features of the problem instance in hand. It would also help to understand what problem features appear to present more difficulty for each method in order to identify directions for further research.

One contribution of this paper is a decomposition method that is adapted to tackle the WSRP with time-dependent activities constraints. The method represents a suitable approach to harness the power of mathematical programming solvers to tackle difficult instances of the WSRP. Another contribution of this paper is a better understanding of the performance by the decomposition method and the tailored constructive heuristic in respect of the problem features. The rest of the paper is organised as follows. Section 2 describes the workforce scheduling and routing problem. Section 3 presents the repeated decomposition with conflict repair method and it also introduces the modification for time-dependent activities constraints. Section 4 presents experimental results from comparing RDCR to the GHI greedy heuristic algorithm. Section 5 concludes the paper.

2   Problem Description

This section describes the workforce scheduling and routing problem with time-dependent activities constraints. The MIP model to solve this problem was originally presented in [11] for a home care crew scheduling scenario. The model is also described in [12,19] and hence not replicated here. As discussed in the introduction, time-dependent activities constraints arise from situations in which visits relate to each other time-wise. Hence, this section also describes the constraints of this type and their formulation.

2.1   Workforce Scheduling and Routing Problem

A network flow model was proposed in [11] for a home care crew scheduling scenario, which is an example of what is called here the workforce scheduling and routing problem (WSRP). That model balances the number of incoming edges and outgoing edges in each node corresponding to a visit location. An edge represents a worker arriving or leaving the visit location. Hence, such balancing means that a worker assigned to a visit must leave the location after performing the task and then move to the next visit location or to the depot. This balancing constraint is applied to each location visit except the depot which is considered as the source and the sink in the network flow model. This same model was also used in [12,19] where other WSRP scenarios with time-dependent activities constraints were tackled using a greedy heuristic algorithm.

The model is a minimisation problem where the objective function is a summation of three main costs. First is the *deployment cost* of assigning each employee to visits. Second is the *preferences cost* of not assigning the most preferred employee to that visit. Third is the *unassigned visit cost* applied when a visit is left unassigned. Each of the three main costs in the objective function is multiplied by weights to give some level of priority to each cost. Here, the values for these weights are set as in [11].

The mixed-integer programming (MIP) model for the problem includes the following constraints. A visit is either assigned to employees or left unassigned. A visit can only be assigned to employees who are qualified to undertake activities associated to the visit. Each path must start from the employee's initial location and end at the final location. The flow conservation constraint guarantees that once employee k arrives to a visit location it then leaves that location in order to form a working path. Visits must start in their starting time window. Assignments of visits to employees must respect the employee's time availability. The time allocated for starting a visit must respect the travel time needed after completing the previous visit. The method presented in this paper has been adapted to tackle time-dependent activities constraints in particular. Such constraints indicate that time-wise dependencies exist between some visits and the specific constraints tackled here are described in more detail next.

2.2   Time-Dependent Activities Constraints

A key difference with previous work described in [16] is that the WSRP scenarios tackled here include a special set of constraints called *time-dependent activities constraints* that establish some inter-dependence between activities. These constraints reduce the flexibility in the assignment of visits to employees because for example, a pair of visits might need to be executed in a given order. There are five constraint types: *overlapping, synchronisation, minimum difference, maximum difference* and *minimum-maximum difference*. A solution that does not comply with the satisfaction of these *time-dependent activities constraints* is considered infeasible.

- *Overlapping* constraint means that the duration of one visit i must extend (partially or entirely) over the duration of another visit j. This constraint is satisfied if the end time of visit i is later than the start time of visit j and also the end time of visit j is later than the start time of visit i.
- *Synchronisation* constraint means that two visits must start at the same time. This constraint is satisfied when the start times of visits i and j are the same.
- *Minimum difference* constraint means that there should be a minimum time between the start time of two visits. This constraint is satisfied when visit j starts at least for a given time units after the start time of visit i.
- *Maximum difference* constraint means that there should be a maximum time between the start time of two visits. This constraint is satisfied when visit j starts at most a given time units after the start time of visit i.
- *Minimum-maximum difference* constraint is a combination of the two previous conditions and it is satisfied when visit j starts at least a time units but not later than another time units after the start time of visit i.

3   The Decomposition Method

This section describes the Repeated Decomposition with Conflict Repair (RDCR) approach used to tackle the WSRP with time-dependent activities constraints. A previous paper [16] presented a method called Geographical Decomposition with Conflict Avoidance (GDCA). In that work, conflicting paths and conflicting assignments as described above were not allowed to happen. However, the existence of time-dependent activities constraints makes it more difficult to just avoid such conflicts when assigning employees to visits. The RDCR method proposed here again seeks to harness the power of exact optimisation solvers by repeatedly decomposing and solving the given problem while also repairing the conflicting paths and conflicting assignments that may arise. The overall RDCR method is presented in Algorithm 1 and outlined next.

The RDCR method takes a WSRP problem denoted by $P = (K, C)$, where C is a set of visits and K is a set of available employees, and applies two main stages. One stage is *problem decomposition* and *sub-problem solving* (lines 2 to 5). The other stage is *conflict repair stage* (lines 6 to 10). The output of RDCR is a solution made by a set of valid paths, each of which is an ordered list of visits assigned to an employee. A valid path is assigned to exactly one employee and does not violate any of the constraints defined in Sect. 2. However, the problem decomposition and sub-problem solving stage may produce conflicting paths, i.e. two or more paths assigned to the same employee. These conflicting paths are then tackled by the conflict repair stage and converted into valid paths. Some visits that were already assigned in the conflicting paths might become unassigned as a result of the repairing process. These unassigned visits are then tackled by repeating the stages of problem decomposition and sub-problem solving followed by conflict repair over some iterations until no more visits can be assigned. The following subsections describe the RDCR method in more detail.

3.1 Problem Decomposition

The problem decomposition (line 2 in Algorithm 1) aims to reduce the size of the feasible region and hence makes possible to tackle the problem with an MIP solver. This process splits the problem into several sub-problems. Each of these sub-problems is made of a subset of employees and visits from the full-size problem but still considering all the types of constraints as in the model described in Sect. 2. Let S be a set of sub-problems $s = (K_s, C_s) \in S$ where K_s and C_s are the subsets of employees and visits respectively for sub-problem s. The outline of the problem decomposition process is shown in Algorithm 2. The two main steps are the visit partition (line 1) and the workforce selection (line 3). These two processes are described in detail next.

Algorithm 1. Repeated Decomposition and Conflict Repair.

Data: Problem $P = (K, C)$ where K is a set of available workforce and C is a set of unassigned visits

Result: {SolutionPaths} FinalSolution

1 **repeat**
2   {Problem} S = ProblemDecomp(K, C);
3   **for** $s \in S$ **do**
4 | sub_sol(s) = **cplex.solve**(s);
5   **end**
6   {Problem} Q = ConflictDetection(sub_sol);
7   FinalSolution.add(NonConflict(sub_sol));
8   **for** $q \in Q$ **do**
9 | cRepair_sol(q) = **cplex.solve**(q);
10   **end**
11   FinalSolution.add(cRepair_sol);
12   Update_UnassignedVisits(C);
13   Update_AvailableWorkforce(K);
14 **until** *No assignment made*;

Algorithm 2. Problem Decomposition.

> **Data**: {Workforce} K, {Visits} C
> **Result**: {Problem} S is a collection of decomposition sub-problems.
> 1 VP = VisitPartition(C);
> 2   **for** $C_i \in VP$ **do**
> 3 ws = WorkforceSelection(K,C_i);
> 4 S.add(subproblem_builder(C_i,ws));
> 5   **end**

Visit Partition. Algorithm 3 shows the steps for the visit partition process. It takes the set of visits C in a full-size problem and produces a partition S consisting of subsets of visits C_i. First, the set of visits C is grouped by location into *visitsList* (since two or more visits might be associated to the same geographical location). Then, each visit c in *visitsList* is allocated to a subset C_i. Basically, the algorithm puts visits that share the same location and visits that are time-dependent into the same subset. The aim of this is that when solving each sub-problem, it becomes easier to enforce the time-dependent activities constraints.

Also, the algorithm observes a maximum size for each subset C_i or sub-problem. This is to have some control over the computational difficulty of solving each sub-problem. As it would be expected, the larger the sub-problem the more computational time required to find an optimal solution or even a feasible one with the MIP solver. However, partitioning into too small sub-problems usually results into solutions of low quality overall. Hence, the sub-problem size is set at 12 visits in our method. However, it is possible for a sub-problem to have more than 12 visits if this means having all activities with the same location and the corresponding time-dependent activities, grouped in the same sub-problem (see line 5 of Algorithm 3).

Workforce Selection. Algorithm 4 shows the steps for the workforce selection process. It takes a subset of visits C_i and the set of employees K to then select a subset of employees ws for the given sub-problem. Basically, for each visit c in C_i the algorithm selects the lowest cost employee w from those employees who are not already allocated to another visit in this same sub-problem (see line 3 of Algorithm 4). That is, an employee w selected for visit c will not be available for another visit in C_i. This process gets a set of employees no larger than $|C_i|$. Note that this method does not generate a partition of the workforce K. This is because although a employee w may be selected for only one visit within subset C_i, such employee w could still be selected for another visit in a different sub-problem, hence potentially generating conflicting paths.

Algorithm 3. Visit Partition Module.

  Data: {Visits} C
  Result: {{Visits}} $VP = \{C_i | i = 1, \ldots, |S|\}$; Partition set of visits
 1 visitsList = OrderByLocation(C);
 2 $i = 0$;
 3 **for** $c \in visitsList$ **do**
 4 **for** $j = 1,\ldots,i$ **do**
 5   **if** $|C_j| <$ *subproblemSize* **or** *c.shareLocation(C_j)* **then**
 6   C_j.add(c);
 7   **if** *c.hasTimeDependent* **then**
 8   Visit c_2 = PairedVisit(c);
 9   C_j.add(c_2);
10   **end**
11   **end**
12 **end**
13 **if** *c.isNotAllocated* **then**
14   $i{=}i{+}1$;
15   C_i.add(c);
16   **if** *c.hasTimeDependent* **then**
17   Visit c_2 = PairedVisit(c);
18   C_i.add(c_2);
19   **end**
20 **end**
21 **end**

3.2 Sub-problem Solving

The problem decomposition process produces a set of sub-problems each with a subset of activities and a subset of selected employees. Each sub-problem is still defined by the MIP model presented in Sect. 2 with its corresponding cost matrix and other relevant parameters. Then, each sub-problem is tackled with the MIP solver (line 4 in Algorithm 1). Solving a sub-problem returns a set of paths. Once the sub-problems are solved there might be conflicting paths, i.e. paths in different sub-problems assigned to the same employee. The conflicting paths require additional steps to resolve the conflict while the valid paths can be used directly. The process to identify and repair such conflicting paths is explained next.

3.3 Conflict Repair

The conflict repair starts by identifying conflicting paths in the solutions to the sub-problems from the problem decomposition. All valid paths are immediately incorporated into the overall solution to the full-size problem. The process to detect conflicting paths is shown in Algorithm 5. It takes all sub-problems solutions and returns the set of conflicting paths Q. Basically, this process searches all sub-problem solutions and identifies all employees who are assigned to two or

---

Algorithm 4. Workforce Selection Module.

    Data: {Visits} C_i, {Workforce} K
    Result: {Workforce} ws
1 **for** $c \in C_i$ **do**
2   Workforce w = bestCostForVisit(K,c,ws);
3   ws.add(w);
4 **end**

---

more paths. It then groups those conflicting paths into sub-problems to repair. This sub-problem to repair has one employee and the set of activities from conflicting paths that belong to that employee.

Algorithm 5. Conflict Path Detection Module

    Data: {SolutionPaths} sub_sol; solutions from solving decomposition
          sub-problems
    Result: {SolutionPaths} Q; Set of conflict paths
1 **for** {$Path$} $s_1 \in sub\_sol$ **do**
2   **for** $Path$ $a_1 \in s_1$ **do**
3   SolutionPaths ConflictPath = null;
4   pathConflicted=false;
5   **for** $s_2 \in sub\_sol$ $|s_2 \neq s_1$ **do**
6   **for** $Path$ $a_2 \in s_2$ **do**
7   **if** $a_1.Employee = a_2.Employee$ **then**
8   ConflictPath.add(a_2);
9   s_2.remove(a_2);
10   pathConflicted = true;
11   **end**
12   **end**
13   **end**
14   **if** $pathConflicted=true$ **then**
15   ConflictPath.add(a_1);
16   s_1.remove(a_1);
17   Q.add(ConflictPath);
18   **end**
19 **end**
20 **end**

In order to repair conflicting paths, the MIP solver tackles the sub-problem to repair which results in a valid path and some unassigned visits. The valid path is incorporated to the solution of the full-size problem. The visits that remain unassigned are tackled by the next iteration of the problem decomposition and sub-problem solving stage followed by the conflict repair stage until no more assignments can be made.

3.4   Tackling Time-Dependent Activities Constraints

As described above, there are five types of time-dependent activities constraints: *overlapping, synchronisation, minimum difference, maximum difference*
and *minimum-maximum difference*. Such time-dependent activities constraints
are usually related to the assignment of two visits. Also, they usually require two
employees, especially the synchronisation and overlapping cases. Hence, these
constraints cannot be enforced by the conflict repair directly because the method
builds a sub-problem to repair based on only one employee. Therefore, modification of the sub-problem to repair is necessary. This is mainly to keep the layout
of assignments when time-dependent conditions are met.

Recall that the problem decomposition and sub-problem solving stage
involves solving sub-problems in which visits share the same location and also
time-dependent visits are grouped in the same sub-problem. Then, as defined
by the MIP model, the solution to a sub-problem satisfies all time-dependent
activities constraints. In order to keep the layout of time-dependent activities,
visits of sub-problems in the conflict repair process require a fixed assigned time
for every time-dependent activity. The fixed time is applied to time window,
i.e. the earliest starting time is equal to the latest starting time for every time-
dependent activity. Once the fixed time restriction is enforced, it affects every
iteration of the process.

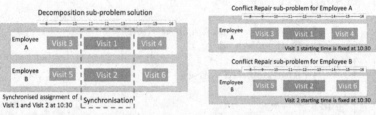

(a) Solution to a decomposition sub-problem   (b) Conflict Repair sub-problems

Fig. 1. Example of tackling time-dependency on synchronised assignments. Sub-figure
(a) shows the solution from solving a decomposition sub-problem. Sub-figure (b) shows
two conflict repair sub-problem solutions. The assigned times from the decomposition
sub-problem solution on visits with time-dependent activities (Visit 1 and Visit 2) are
carried on to the later stage of the process. The time windows of Visit 1 and Visit 2
are fixed to the same value when preparing conflict repair sub-problem. Fixed starting
time is enforced on Visit 1 and Visit 2 until both of them are incorporated into the
final solution or the iterative process is terminated.

Figure 1 shows an example of how the modification works on a synchronisation constraint. With reference to the figure, suppose that visit 1 and visit
2 must be synchronised. Because visit 1 and visit 2 are time-dependent, they
are grouped into the same decomposition sub-problem. The decomposition sub-
problem is solved which gives paths for employee A and employee B, as shown

in Fig. 1(a). From that sub-figure, visit 1 and visit 2 are assigned to employee A and employee B, respectively. Both visits have their starting time set at 10:30. Suppose that both paths of employee A and employee B need to be repaired. At this stage, the time-dependent modification is applied. It overrides the time window of both visits and sets them to 10:30. Here, there are two sub-problems to repair, presented in Fig. 1(b). Recall that a sub-problem to repair is defined based on an employee who has conflicting paths. Both sub-problems apply the new time window values forcing the start time of visit 1 and visit 2 to 10:30. The new time window is enforced until both visits are assigned to the final solution or the iterative process is terminated.

In the same way, the modification explained above tackles the other types of time-dependent constraints. The time-dependent visits are grouped in the same sub-problem and the solution of this part satisfies time-dependent activities constraints in the decomposition step. The time-dependent modification also applies when the time-dependent visit needs to be repaired. The modification replaces the time window of the visit by a fixed time given by the decomposition step. Then, this modification ensures that a solution that has gone through the conflict repair will satisfy the time-dependent activities constraints.

4   Experiments and Results

This section describes the experiments carried out to compare the proposed RDCR method to the greedy heuristic (GHI) in [12] and better understand the success of each method according to features of the problem instances.

4.1   WSRP Instances Set

The RDCR method was applied to the set of WSRP instances presented in [10,12]. Those problem instances were generated by adapting several WSRP from the literature. The instances are categorised in four groups: Sec, Sol, HHC and Mov. The Sec group contains instances from a security guards patrolling scenario [20]. The Sol group are instances adapted from the Solomon dataset [21]. The HHC group are instances from a home health care scenario [11]. Finally, the Mov group originates from instances of the vehicle routing problem with time windows [22]. The total number of instances accumulated in these four groups is 374.

4.2   Overview of Greedy Heuristic GHI

A greedy constructive heuristic tailored for the WSRP with time-dependent activities constraints was proposed in [12]. The algorithm starts by sorting visits according to some criteria such as visit duration, maximum finish time, maximum start time, etc. Then, it selects the first unassigned visit in the list and applies an *assignment process*. For each visit c, the *assignment process* selects all candidate employees who can undertake visit c (considering required skills and availability).

If the number of candidate employees is less than the number of employees required for visit c, this visit is left unassigned. If visit c is assigned, visits that are dependent on visit c are processed. These dependent visits c' jump ahead in the *assignment process* and are themselves processed in the same way (i.e. processing other visits dependent on c'). The GHI stops when the unallocated list is empty and then returns the solution.

4.3 Computational Results

The proposed RDCR method was applied to the 374 instances and the obtained solutions were compared to the results reported by the greedy heuristic (GHI).

First, the related-samples Wilcoxon Signed Rank Test [23] was applied to examine the differences between the two algorithms, GHI and RDCR. The significant level of the statistical test was set at $\alpha = 0.05$. Results of this statistical test using SPSS are shown in Table 1 showing that RDCR produced better solutions for 209 out of the 374 instances. However, there was no statistical significant difference on the solution quality between the two methods.

Figures 2 and 3 compare the number of best solutions found by each of the two methods and the average relative gap to the best known solutions. In these figures, results are grouped by dataset. Note that the relative gap is calculated by $\Delta = |z - z^b|/|z^b|$ where z represents an objective value of a solution and z^b

Table 1. Statistical result from related-samples Wilcoxon signed rank test provided by SPSS.

Total N	374
# of (RDCR < GHI), RDCR is better than RDCR	209
# of (RDCR > GHI), GHI is better than GHI	165
Test statistic	37,806
Standard error	2,092
Standardized test statistic	1.311
Asymp. Sig. (2-sided test)	.190

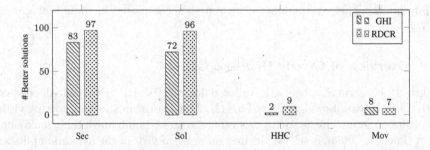

Fig. 2. Number of best solutions obtained by GHI and RDCR for each dataset.

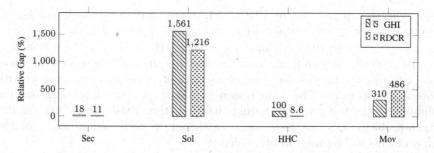

Fig. 3. Average relative gap (relative to the best known solution) obtained by GHI and RDCR. The lower the bar the better, i.e. the closer to the average best known solution.

is an objective value of the best known solution. Regarding the number of best solutions, RDCR produced better results than GHI on three datasets: Sec, Sol and HHC. Results also show that RDCR found lower values of average relative gap on the same three datasets. On the Mov dataset, GHI performed better in terms of number of best solutions and average relative gap.

On datasets Sec and Sol, RDCR found slightly better results than GHI as shown by the number of best solutions and the average relative gap. In dataset Sec, RDCR and GHI gave 11% and 18% of average relative gap respectively. This indicates that both algorithms provide good solution quality compared to the best known solution. On the other hand, both RDCR and GHI produced 1,216% and 1,561% respectively for the average relative gap to the best known solution in dataset Sol. This implies that both algorithms failed to find solutions that are of competitive quality to the best known solution, but both algorithms are competitive between them. It can be seen that instances in this Sol dataset are particularly difficult as neither the GHI heuristic nor the RDCR decomposition technique could produce solutions of similar quality to the best known solution.

On dataset HHC, the average relative gap of RDCR is much lower than the average gap of GHI. The results show that RDCR has 8.6% relative gap while GHI has 100%. For the HHC instances, RDCR found the best known solution for 9 instances and GHI found the best known solution for the other 2 instances. For these two instances, average relative gap of RDCR is 47%. However, in the 9 best solutions of RDCR, average gap of GHI is 109%. A closer look at the Sol dataset showed that these instances have priority levels defined for the visits. It turns out that GHI does not have sorting parameters to support such priority for visits because the algorithms sorting parameters focus on the time and duration of visits. On the other hand, RDCR implemented priority for visits within the MIP model. This could be the reason that explains the better results obtained by RDCR on this particular dataset.

On dataset Mov, GHI gives better performance. GHI delivers 8 better solutions (7 best known) from 15 instances while RDCR gives 7 better solutions (4 best known). The average relative gap of GHI is 310% which is less than the

486% relative gap provided by RDCR. There are 5 instances which best known solution is given by the mathematical programming solver. For these, the average relative gaps to the best known given by GHI is 315% and by RDCR is 36% respectively. It was found that the decomposition method does not show good performance on this particular Mov dataset, especially on instances with more than 150 visits. The main reason is that the solver cannot find optimal solutions to the sub-problems within the given time limit. Therefore, the size of sub-problems in these Mov instances should be decreased to allow for the sub-problems to be solved to optimality.

Figure 4 shows the cumulative distribution of RDCR and GHI solutions over the relative gap. It shows the number of solutions which have a relative gap to the best known less than the corresponding value in the X-axis. Note that 0% relative

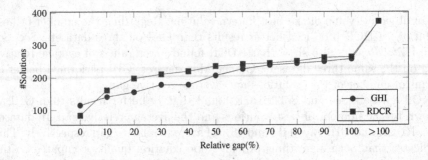

Fig. 4. Cumulative distribution of GHI and RDCR solution over the relative gap.

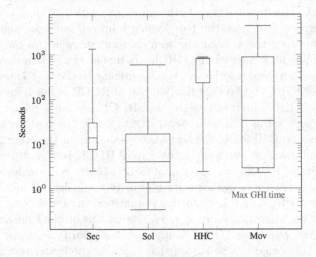

Fig. 5. Box and Whisker plots showing the distribution of computational time in seconds spent by RCDR for each group of instances. The wider the box the larger the number of instances in the group. The orange straight line presents the upper limit in the computational time spent by GHI (fixed to 1 second). The Y-axis is in logarithmic scale.

gap refers to the best known solution. For this case, GHI provides 115 best known solutions which is better than RDCR which provides 84 best solutions. This is represented by the two leftmost points in the figure. However, from the value of 10% relative gap onwards, RDCR delivers larger number of solutions than GHI. Overall, apart from the overall number of best known solutions, RDCR provides higher number (or equal) of solutions than GHI for different values of relative gap. For example, if the solution acceptance rate is set at 50% relative gap, RDCR produces 236 solutions of this quality while GHI produces 207. Hence, RDCR delivers overall more solutions with acceptance rate up to 100% gap to the best known.

Figure 5 shows the distribution of computational time spent by the proposed RDCR method when solving the WSRP instances considered here. These results show that RDCR spends more computational time on most of the HHC instances with an overall average time spent on each instance of 2.4 min. Note that the highest computational time observed in these experiments is less than 74 min. On the other hand, the computational time spent by GHI is much shorter, taking less than one second on each instance. Therefore, GHI is clearly superior to RDCR in terms of computational time.

4.4 Performance According to Problem Difficulty

This part seeks to better understand the performance of the two algorithms GHI and RDCR. For this, a more detailed analysis is conducted of the instances in which each of the algorithms performs better than the other one. Then, the problem features are analysed in detail in order to unveil any conditions under which each of the algorithms appears to performs particularly well.

Table 2 presents the main characteristics of the problem instances in three groups. *Set All* has the 374 instances. *Set GHI* has all problem instances in which GHI produced better solutions than RDCR. *Set RDCR* has all problem instances in which RDCR produced better solutions than GHI. The table shows average and standard deviation values for 8 problem characteristics: the number of employees (#Emp), the number of visits (#Visit), visit duration (Visit-Dur), the number of time-dependent activities (#TimeDep), employee-visit ratio (Emp/Visit), employee available hours (EmpHours), average visit time window (VisitWindow), and planning horizon (Horizon). These values are presented in the table in the format Mean $\pm$ SD. Those problem characteristics for which there is a statistical significant difference between *Set GHI* and *Set RDCR* (using t-test at significance level $\alpha = .05$) are marked with *.

It seems obvious to relate the difficulty of a particular problem instance to it size, which can be measured by the number of employees and the number of visits. It could also be assumed that the length of the planning horizon might have some influence on the difficulty of the problem in hand, although perhaps to a lesser extent than the number of employees and visits. However, the analysis presented here seeks to identify other problem characteristics that might have an effect of the difficulty of the instances when tackled by each of the algorithms RDCR and GHI. For example, it can be argued that having visits with longer

Table 2. Summary of the problem features for different groups of problem instances. The *Set All* includes all instances. The *Set GHI* includes the instances in which GHI produces better solutions than RDCR. The *Set RDCR* includes the instances in which RDCR produces better solutions than GHI. Values are displayed in the format mean ± std. dev.

# Instances in group	*Set all* 374	*Set GHI* 165	*Set RDCR* 209
Problem size			
#Emp*	22.5 ± 22.55	31.37 ± 27.86	15.49 ± 16.45
#Visit*	87.22 ± 53.23	117.6 ± 54.65	63.26 ± 37.83
Characteristics on visits and employees			
VisitDur*	214.3 ± 198.8	254.6 ± 221.6	182.4 ± 173.4
#TimeDep*	13.95 ± 10.03	18.87 ± 10.63	10.37 ± 7.91
Emp/Visit*	1.164 ± 0.072	1.156 ± 0.074	1.172 ± 0.079
EmpHours	20.8 ± 11.93	21.66 ± 11.45	20.11 ± 12.31
VisitWindow	392.9 ± 297.6	406.9 ± 325.9	381.9 ± 274.2
Horizon	1248 ± 715.9	1300 ± 687.2	1207 ± 738.4

* indicates statistical significant difference using t-test at significance level $\alpha = .05$.

duration or large number of time-dependent activities could make the problem instance more difficult to solve because of the higher likelihood of time conflicts arising. In contrast, the difficulty could decrease for a problem instance that has higher employee to visit ratio (i.e. more workers to choose from), longer employee working hours or wider visit time windows (i.e. more flexibility for the assignment of visits).

Considering the above, it seems from Table 2 that instances in *Set RDCR* are less difficult than those in *Set GHI*. In respect of the problem size, instances in *Set RDCR* are on average smaller than those in *Set GHI*, on the number of employees (#Emp) and also the number of visits (#Visit). In addition, instances in *Set RDCR* have shorter visit duration (VisitDur) and lower number of time-dependent activities (#TimeDep) than instances in *Set GHI*. Moreover, note that although the averages values of employee-visit ratio (Emp/Visit) are very similar for sets *Set RDCR* and *Set GHI*, the difference is still statistically significant. The differences between the two sets in respect of the remaining three problem characteristics, employee available hours (EmpHours), visit time window (VisitWindow) and planning horizon (Horizon) were found to be not statistically significant.

Then, from the above analysis it can be argued that the RDCR approach performs better than GHI on instances of lower difficulty level. However, establishing the boundary between lower and higher difficulty is not so clear given the overlap in values for the 8 problem characteristics between *Set RDCR* and *Set GHI*. Hence, the proposal here is to recommend the use of RDCR for instances with less than 22.5 employees and less than 87 visits (the average values considering all 374 instances), and the use of GHI otherwise. This recommendation can be used as a first step for choosing between RDCR and GHI.

4.5   Performance on Producing Acceptable Solutions

The previous subsection sought to identify a boundary in problem difficulty between those instances in which each of the methods RDCR and GHI performs better than the other one. This subsection seeks to identify instances for which both algorithms can deliver acceptable solutions. For this, a solution that has a relative gap of at most 100% with respect to the best known solution is considered acceptable, otherwise it is labelled unacceptable.

The first part of the analysis splits the problem instances into two groups. The group *Accept Heur* has instances for which an acceptable solution was found by at least one of the two heuristic algorithms RDCR and GHI. The group *Reject Heur* has instances for which none of RDCR or GHI delivers an acceptable solution. Basically, this analysis seeks to identify a boundary in problem difficulty for which the methods RDCR and GHI can perform better than an exact solver. Table 3 shows the problem characteristics for the two groups *Accept Heur* and *Reject Heur*. As before, each row shows the average and standard deviation values for each of 8 problem characteristics. Those problems characteristics for which there is a statistical significant difference between the two groups (using t-test at significance level $\alpha = .05$) are marked with *.

The results in Table 3 show that there are significant differences between the groups *Accept Heur* and *Reject Heur* on six problem characteristics. That is,

Table 3. Summary of the problem features for different groups of problem instances. The group *Accept Heur* includes instances for which an acceptable solution was found by at least one of the two heuristic algorithms RDCR and GHI. The group *Reject Heur* includes instances for which none of RDCR or GHI delivers an acceptable solution. Values are displayed in the format mean ± std. dev.

# Instances in group	Reject heur 79	Accept heur 295
Problem size		
#Emp*	9.911 ± 4.249	25.87 ± 25.40
#Visit*	49.39 ± 21.17	97.34 ± 54.75
Characteristics on visits and employees		
VisitDur*	43.91 ± 53.54	259.89 ± 199.09
#TimeDep*	6.32 ± 3.11	15.99 ± 10.27
Emp/Visit	1.168 ± 0.048	1.164 ± 0.078
EmpHour*	18.02 ± 13.69	21.54 ± 11.34
VisitWindow	345.60 ± 387.58	405.58 ± 268.43
Horizon*	1081.41 ± 821.62	1292.61 ± 608.71

* indicates statistical significant difference using t-test at significance level $\alpha = .05$.

the group *Accept Heur* shows higher mean values than the group *Reject Heur* for the number of employees (#Emp), the number of visits (#Visit), visit duration (VisitDur), the number of time-dependent activities (#TimeDep), employee available hours (EmpHours), and planning horizon (Horizon). These results indicate that GHI and RDCR do not provide acceptable solutions on the smaller instances with around 10 employees and 50 visits. However, these algorithms do well on the larger instances with around 26 employees and 97 visits. This is because the exact solver performs very well on the smaller instances but not so well when the problem size grows. Hence, the proposal here is to recommend the use of the exact solver for problems with less than 15 employees and 70 visits. For larger problem instances the solver may spend too long time finding solutions hence it is better to use GHI or RDCR considering the recommendation in the previous subsection.

The second part of the analysis analysis splits the 295 problem instances from the group *Accept Heur* into groups according to whether the particular method GHI or RDCR produces acceptable solutions or not. As before, a solution that has a relative gap of at most 100% with respect to the best known solution is considered acceptable, otherwise it is labelled unacceptable. Table 4 shows the split for method GHI into groups *Accept GHI* with 258 instances and *Reject GHI* with 37 instances. There are significant differences between the two groups on four characteristics: the number of employees (#Emp), the number of visits (#Visit), visit duration (VisitDur) and the number of time-dependent activities (#TimeDep) with larger values for the group *Accept GHI*. These results confirm that GHI provides acceptable solutions on the larger instances but it struggles to produce acceptable solutions for some smaller instances.

Table 4. Summary of the problem features for different groups of problem instances. The group *Accept GHI* includes instances for which an acceptable solution was found by algorithm GHI, otherwise the instance is included in group *Reject GHI*. Values are displayed in the format mean ± std. dev.

# Instances in group	*Reject GHI* 37	*Accept GHI* 258
Problem size		
#Emp*	15.97 ± 31.23	27.29 ± 24.19
#Visit*	54.86 ± 49.79	103.43 ± 52.77
Characteristics on visits and employees		
VisitDur*	42.06 ± 53.77	291.13 ± 192.69
#TimeDep*	8.00 ± 10.80	17.14 ± 9.69
Emp/Visit	1.1471 ± 0.07873	1.166 ± 0.0779
EmpHour	23.19 ± 20.27	21.30 ± 9.44
VisitWindow	462.90 ± 468.99	397.36 ± 226.01
Horizon	1391.78 ± 1216.20	1278.38 ± 566.81

* indicates statistical significant difference using t-test at significance level $\alpha = .05$.

Table 5. Summary of the problem features for different groups of problem instances. The group *Accept RDCR* includes instances for which an acceptable solution was found by algorithm RDCR, otherwise the instance is included in group *Reject RDCR*. Values are displayed in the format mean ± std. dev.

# Instances in group	*Reject RDCR* 31	*Accept RDCR* 264
Problem size		
#Emp*	46.74 ± 54.91	23.42 ± 17.88
#Visit	155.6 ± 59.01	95.20 ± 53.95
Characteristics on visits and employees		
VisitDur*	41.19 ± 62.94	285.57 ± 193.80
#TimeDep	15.35 ± 7.38	16.07 ± 10.57
Emp/Visit*	1.077 ± 0.0654	1.174 ± 0.0731
EmpHour	20.82 ± 16.58	21.62 ± 10.60
VisitWindow	407.11 ± 453.92	405.40 ± 238.84
Horizon	1249.41 ± 995.01	1297.68 ± 636.26

* indicates statistical significant difference using t-test at significance level $\alpha = .05$.

Table 5 shows the split for method RDCR into groups *Accept RDCR* with 264 instances and *Reject RDCR* with 31 instances. There are significant differences between the two groups on three characteristics: the number of employees (#Emp), visit duration (VisitDur) and employee-visit ratio (Emp/Visit). The size of instances in group *Accept RDCR* seems smaller than in group *Reject RDCR* as given by #Emp and #Visit, although only for #Emp the difference is significant. Instances in the group *Reject RDCR* have shorter visit duration and lower employee-visit ratio. A problem instance could become more difficult to solve if there are less workers to be assigned to visits. These results confirm that the performance of RDCR on providing acceptable solutions suffers as the size of the problem grows.

From the above analysis on producing acceptable solutions, some recommendations can be drawn in respect of what type of approach to use according to the problem size. Table 6 shows the type of approach recommended according to

Table 6. Type of approach recommended according to the problem size and number of instances in each size class.

Algorithm	Exact method	RDCR	Heuristic	GHI
#Instance	79	37	227	31
Problem size	Very small	Small	Medium	Large
Average #Emp	9.91	15.97	23.42 - 27.29	49.74
Average #Visit	49.39	54.86	95.20 - 103.43	155.6

the problem size and number of instances in each size class. The first row of the table shows the suggested algorithm for each size class, Heuristic refers to either GHI or RDCR. For each size class, the table shows the number of instances (#Instance), the problem size label, the average number of employees (Average #Emp) and the average number of visits (Average #Visit). It is suggested that to use the exact method to solve very small instances, to use RDCR to solve small and medium instances and to use GHI to solve medium and large instances. The problem size class with the largest number of instances is the medium class for which the two heuristic algorithms, GHI and RDCR, find acceptable solutions. These recommendations in Table 6 were drawn from looking at the reject groups in Tables 3, 4 and 5. Both GHI and RDCR do not perform well when solving small instances, given that group *Reject Heur* in Table 3 has the smallest average problem size. RDCR should be used for instances larger than those in group *Reject Heur*, Table 4 shows that the *Reject GHI* group has average problem size larger than the *Reject Heur* group and smaller than the *Reject RDCR* group. GHI tends to be effective in the largest instance group, it can be seen from Table 5 that the *Reject RDCR* group has the largest average problem size compared to the *Reject GHI* group and *Reject Heur*. However, both RDCR and GHI have similar performance as their acceptable solutions are similar in number.

5   Conclusion

This paper presented a decomposition method for mixed integer programming to solve instances of the workforce scheduling and routing problem (WSRP) with time dependent activities constraints. The method uses heuristic partition and selection to split a problem into sub-problems. A sub-problem solution gives a path or sequence of visits for each employee. Each sub-problem is individually solved by the MIP solver. Within a sub-problem solution or path, all constraints are satisfied. Paths may conflict with paths from other sub-problems, i.e. two or more different sequences of visits but assigned to the same employee. This can be fixed by a conflict repair process. However, conflict repair requires modification to support time-dependent activities constraints since the repairing process may rearrange assignment time. Thus, the modification maintains the layout of time-dependent activities by fixing the assigned time of the time-dependent activities. Therefore, the solution from conflict repair does not violate any constraints.

The proposed RDCR approach is applied to solve four WSRP scenarios with a total of 374 instances. The experimental results showed that RDCR is able to find better solutions than the GHI heuristic for 209 out of the 374 instances. However, the statistical test showed that RDCR does not perform significantly different to the deterministic greedy heuristic (GHI). RDCR showed better performance on three out of four datasets. The computational time required to solve a problem instance with RDCR ranged from less than a second to 74 min. The average computational time was under 3 min. Overall, the proposed RDCR with time-dependent modification is able to effectively solve WSRP instances with time-dependent activities constraints. The method found competitive feasible solutions to every instance and within reasonable computational time.

The paper also conducts a study to investigate the performance of RDCR in respect of some problem features related to the problem size. The analysis has shown that RDCR provides better solutions particularly in smaller instances. Hence, instances with less than 87 visits and less than 22 workers should be tackled by RDCR to obtain higher quality solutions. Furthermore, another aim of the study was to determine the class of problem size that can be more effectively tackled with the heuristic approaches RDCR and GHI. For this, acceptable solutions are considered to be those that have a relative gap of no more than 100% with respect to the best known solution. The analysis revealed that RDCR and GHI work effectively in a wide range of problem sizes. The GHI method appears to be less effective on smaller instances while RDCR appears to be less effective on larger instances. Therefore, in order to produce acceptable solutions as defined here, it is recommended to use an exact solver for very small instances, to use RDCR for small and medium instances and to use GHI for medium and large instances.

As future work it is suggested to improve the computational time of the proposed RDCR approach. Such improvement might be achieved by applying different methods to partition the set of visits or by using more effective workforce selection rules. Also, determining the right sub-problem size could be interesting as it could help to balance solution quality and time spent on computation.

Acknowledgements. The authors are grateful for access to the University of Nottingham High Performance Computing Facility. Also, the first author thanks the DPST Thailand for partial financial support of this research.

References

1. Hiermann, G., Prandtstetter, M., Rendl, A., Puchinger, J., Raidl, G.R.: Metaheuristics for solving a multimodal home-healthcare scheduling problem. CEJOR **23**, 89–113 (2015)
2. Borsani, V., Andrea, M., Giacomo, B., Francesco, S.: A home care scheduling model for human resources. In: 2006 International Conference on Service Systems and Service Management, pp. 449–454 (2006)
3. Eveborn, P., Flisberg, P., Rönnqvist, M.: Laps care-an operational system for staff planning of home care. Eur. J. Oper. Res. **171**, 962–976 (2006)
4. Angelis, V.D.: Planning home assistance for AIDS patients in the city of Rome, Italy. Interfaces **28**, 75–83 (1998)
5. Leigh, J., Jackson, L., Dunnett, S.: Police officer dynamic positioning for incident response and community presence. In: Proceedings of the 5th International Conference on Operations Research and Enterprise Systems (ICORES 2016), pp. 261–270 (2016)
6. Misir, M., Smet, P.V.B.G.: An analysis of generalised heuristics for vehicle routing and personnel rostering problems. J. Oper. Res. Soc. **66**, 858–870 (2015)
7. Jean-François, C., Gilbert, L., Federico, P., Stefan, R.: Scheduling technicians and tasks in a telecommunications company. J. Sched. **13**, 393–409 (2010)

8. Laugier, A., Anne-marie, B., Telecom, R. F.: Technicians and interventions scheduling for telecommunications. Francetelecom, 1–7 (2006)
9. Lesaint, D., Voudouris, C., Azarmi, N.: Dynamic workforce scheduling for British telecommunications plc. Interfaces **30**, 45–56 (2000)
10. Castillo-Salazar, J., Landa-Silva, D., Qu, R.: Workforce scheduling and routing problems: literature survey and computational study. Ann. Oper. Res. (2014)
11. Rasmussen, M.S., Justesen, T., Dohn, A., Larsen, J.: The home care crew scheduling problem: preference-based visit clustering and temporal dependencies. Eur. J. Oper. Res. **219**, 598–610 (2012)
12. Castillo-Salazar, J.A., Landa-Silva, D., Qu, R.: A greedy heuristic for workforce scheduling and routing with time-dependent activities constraints. In: Proceedings of the 4th International Conference on Operations Research and Enterprise Systems (ICORES 2015) (2015)
13. Mankowska, D., Meisel, F., Bierwirth, C.: The home health care routing and scheduling problem with interdependent services. Health Care Manage. Sci. **17**, 15–30 (2014)
14. Xu, J., Chiu, S.: Effective heuristic procedures for a field technician scheduling problem. J. Heuristics **7**, 495–509 (2001)
15. Dantzig, G.B., Wolfe, P.: Decomposition principle for linear programs. Oper. Res. **8**, 101–111 (1960)
16. Laesanklang, W., Landa-Silva, D., Castillo-Salazar, J.A.: Mixed integer programming with decomposition to solve a workforce scheduling and routing problem. In: Proceedings of the 4th International Conference on Operations Research and Enterprise Systems (ICORES 2015), pp. 283–293 (2015)
17. Laesanklang, W., Pinheiro, R.L., Algethami, H., Landa-Silva, D.: Extended decomposition for mixed integer programming to solve a workforce scheduling and routing problem. In: Werra, D., Parlier, G.H., Vitoriano, B. (eds.) ICORES 2015. CCIS, vol. 577, pp. 191–211. Springer, Cham (2015). doi:10.1007/978-3-319-27680-9_12
18. Reimann, M., Doerner, K., Hartl, R.F.: D-Ants: savings based ants divide and conquer the vehicle routing problem. Comput. Oper. Res. **31**, 563–591 (2004)
19. Laesanklang, W., Landa-Silva, D., Castillo-Salazar, J.A.: Mixed integer programming with decomposition for workforce scheduling and routing with time-dependent activities constraints. In: Proceedings of 5th the International Conference on Operations Research and Enterprise Systems, pp. 330–339 (2016)
20. Misir, M., Smet, P., Verbeeck, K., Vanden Berghe, G.: Security personnel routing and rostering: a hyper-heuristic approach. In: Proceedings of the 3rd International Conference on Applied Operational Research, ICAOR 2011, pp. 193–205 (2011)
21. Solomon, M.M.: Algorithms for the vehicle routing and scheduling problem with time window constraints. Oper. Res. **35** (1987)
22. Castro-Gutierrez, J., Landa-Silva, D., Moreno, P.J.: Nature of real-world multi-objective vehicle routing with evolutionary algorithms. In: 2011 IEEE International Conference on Systems, Man, and Cybernetics (SMC), pp. 257–264 (2011)
23. Field, A.: Discovering Statistics Using IBM SPSS Statistics, 4th edn. SAGE Publication Ltd., London (2013)

Author Index

Printed in the United States
by Bookmasters

Printed in the United States
By Bookmasters